Lecture Notes in Physics

Edited by H. Araki, Kyoto, J. Ehlers, München, K. Hepp, Zürich
R. Kippenhahn, München, D. Ruelle, Bures-sur-Yvette
H.A. Weidenmüller, Heidelberg, J. Wess, Karlsruhe and J. Zittartz, Köln
Managing Editor: W. Beiglböck

340

M. Peshkin
A. Tonomura

The Aharonov-Bohm Effect

Springer-Verlag
Berlin Heidelberg GmbH

Authors

M. Peshkin
Physics Division, Argonne National Laboratory
9700 South Cass Avenue, Argonne, IL 60439–4843, USA

A. Tonomura
Advanced Research Laboratory, Hitachi, Ltd.
Kokubunji, Tokyo 185, Japan

ISBN 978-3-662-13727-7   ISBN 978-3-540-46661-1 (eBook)
DOI 10.1007/978-3-540-46661-1

2158/3140-543210 – Printed on acid-free paper

PREFACE

The Aharonov-Bohm effect is a quantum-mechanical phenomenon wherein the motion of a charged particle is influenced by the existence of electromagnetic fields in regions which the particle does not enter. The influence of those remote fields is carried in the theory by the four-vector potential, which appears in the Hamiltonian and therefore in the Schroedinger equation. This phenomenon is counter-intuitive for many physicists because it has no classical analog and because it challenges the conventional perception that the physical quantities in electromagnetism are carried by the local Maxwell fields, not by the potentials. If confirmed by experiment, the Aharonov-Bohm effect raises fundamental questions about locality in quantum mechanics and in the electromagnetic interaction. If refuted, it requires either a drastic revision of supposedly fundamental ideas about quantum mechanics or a demonstration that some error has been made and standard quantum mechanics does not really imply the existence of the Aharonov-Bohm effect. Either way, it addresses questions about quantum mechanics in a multiply-connected region that have been recognized as options in the theory but that have not been investigated by sufficiently sensitive experiments in the past.

These considerations have stimulated a lively theoretical and experimental effort for the past thirty years, with almost every possibility represented and with much debate about the status of experiments and of calculations. Now the main issues appear to be clear. Almost everyone agrees that the Aharonov-Bohm effect is confirmed by experiment and that it is a genuine feature of standard quantum mechanics. In this treatise, we give an account of the experiments and of the fundamental theoretical questions that they illuminate. Our intended audience includes students who have learned the elements of quantum mechanics and physicists interested in fundamental questions. We have had great pleasure in exploring the new physics introduced by the Aharonov-Bohm effect and the old physics on which it sheds a new light, and we hope to transmit that pleasure to the reader.

In Part One, Peshkin introduces the Aharonov-Bohm effect under the assumptions of standard quantum theory. The emphasis here is on the fundamental questions which are addressed directly by the experimental results and on other elementary aspects of the the theory that appear especially clearly in the present context. In Part Two, Tonomura gives a brief review of the diverse ideas that have been put forward during the protracted debate about the reality and meaning of the Aharonov-Bohm effect, followed by a detailed account of the experiments which finally settled most of those questions. The review is intended to give a flavor of the controversy rather than to do justice to every point of view.

We thank Professor Huzihiro Araki of Kyoto University and Professor Mikio Namiki of Waseda University for their kind recommendations and for arranging to have this treatise published in Lecture Notes in Physics. Peshkin is indebted to Professor Harry J. Lipkin of the Weizmann Institute for contributions that cannot even be identified after decades of close collaboration. Tonomura thanks Professor Hiroshi Ezawa of Gakushuin University and Professor Mark P. Silverman of Trinity College, Hartford, for useful discussions during the preparation of this manuscript.

March, 1989

Murray Peshkin

Akira Tonomura

CONTENTS

PART ONE: THEORY (MURRAY PESHKIN)

PART TWO: EXPERIMENT (AKIRA TONOMURA)

The Aharonov-Bohm Effect

Part One: Theory

Murray Peshkin

Argonne National Laboratory, Argonne
Illinois 60439, USA

1. INTRODUCTION

According to standard quantum mechanics, the motion of a charged particle can sometimes be influenced by electromagnetic fields in regions from which the particle is rigorously excluded [1,2]. This phenomenon has come to be called the Aharonov-Bohm effect (AB effect), after the seminal 1959 paper entitled "Significance of Electromagnetic Potentials in the Quantum Theory," by Y. Aharonov and D. Bohm [2]. What AB effect teaches us about the significance of the electromagnetic potentials has since been discussed from several points of view [3-8], on the assumption that standard quantum mechanics is indeed a correct description of nature.

However, the discussion has gone much further, driven in part by some physicists' disbelief in the possibility of observable effects of fields confined to excluded regions, in part by the opportunity to test quantum mechanics in a new regime, and in part by the opportunity to understand the workings of the theory in a new way. AB effect has been the subject of more than three hundred journal articles in the past thirty years. Calculations purportedly based on standard quantum mechanics have been interpreted as showing that AB effect does not exist in the theory, that Aharonov and Bohm are simply in error [9-12]. The Ehrenfest theorem has been invoked to prove that something is wrong somewhere: with no forces, a particle or a wave packet cannot be deflected. Modified versions of quantum mechanics which do not exhibit AB effect but are claimed to share the tested predictions of the standard theory have been put forward [13]. Classical calculations have been interpreted to show that the AB effect does not actually describe a particle moving in a field-free region, that the interaction with the source of the fields in the excluded region results in an induced classical force on the particle [14,15]. All these ideas have been refuted by theoretical analyses which support the conclusions and sometimes the interpretations of Aharonov and Bohm [16].

The experimental quantization of the fluxoid in superconducting rings and in Josephson junctions has been interpreted as an experimental confirmation of AB effect [17]. Interference experiments on electron beams have been carried out to provide more direct confirmation, with increasing precision and especially with increasing control of stray fields that might obscure the implications of the experiments [18-20].

There have also been extensions of the original idea. AB effect with a non-Abelian gauge field replacing the electromagnetic field has been described in theory [7,8,21,22], although the chance for a feasible experiment seems remote. The theoretically possible existence of objects consisting of electrons bound to magnetic flux lines, with unusual spins and possibly unusual statistics, has been suggested [23]. Practical use of AB effect to study the quantum properties of mesoscopic normal conductors is being developed [24,25]. AB effect has also been used in a novel experiment to measure the charge of the neutron [26], and very recent experiments have revealed the structure of flux lines in superconductors [27].

Many of the theoretical disagreements have been between authors who claimed or implied that they started from the same assumptions. Others result from incompleteness of the standard assumptions when the domain of a particle is a multiply connected region, as it always is in AB effect. Some of those authors who denied the existence of

AB effect in the theory have challenged the positive experimental results by questioning the experimenters' claimed elimination of error due to stray fields.

Now the decisive experiment has been done [20]. It confirms the predictions of Aharonov and Bohm with exquisite precision and control of the stray field problem. This treatise is intended to serve as a preface to and appreciation of the following one by A. Tonomura, in which he describes both his experiments and the earlier experimental efforts. Most of what I present here is not substantially new. My purpose is twofold: to introduce the experiment by outlining the theoretical ideas that it tests, and to discuss the fundamental issues in physics that have been addressed by the theory and the experiment. Almost all of the discussion assumes nothing more than nonrelativistic quantum mechanics based on the Schroedinger equation or on algebraic consequences of the commutation relations. Much of it relies on only a few general properties of the theory. I believe that all the central issues are best illuminated by this minimalist approach, which emphasizes that AB effect is deeply involved with the most primitive and general features of quantum theory.

I do not discuss the many theoretical ideas which have been advanced to remove AB effect from the theory because the experiments have now negated those attempts, and also because I am unaware of any such idea that seems to me to have led to a viable theoretical structure even if one ignores the experiments. Obviously, there are other points of view. The review by Olario and Popescu [28] contains a good guide to all of them. Its comprehensive bibliography is updated in Tonomura's paper, which also gives a concise review of some of the disparate theoretical claims from the more neutral perspective of an experimenter who will put them to the test.

2. WHAT IS THE AHARONOV-BOHM EFFECT?

The concept was introduced in Ref. [2] as follows: Consider the interference experiment illustrated in Fig. 1. Electrons enter from the left and the beam is split coherently in a two-arm interferometer. In principle, any change in the relative phase between the beams in the two arms can be observed as a shift in the interference pattern when the two beams are reunited at the right.

In the magnetic version of AB effect, a stationary magnetic field is introduced in the region between the two beams, as in Fig. 1a. The electrons are forever rigorously excluded from that region by some baffles. The return magnetic flux is made to avoid the regions where the electrons are permitted. The Hamiltonian H and the time-independent wave function $\psi(x)$ are given by[*]

$$H = (1/2m)[-i\hbar\nabla+(e/c)A_e]^2-eV_0(x) \tag{2.1}$$

$$\psi(x) = \psi_0(x) \exp\{-iS(x)/\hbar\} , \tag{2.2}$$

---

[*]I use Gaussian units. e represents the absolute value of the electron's charge.

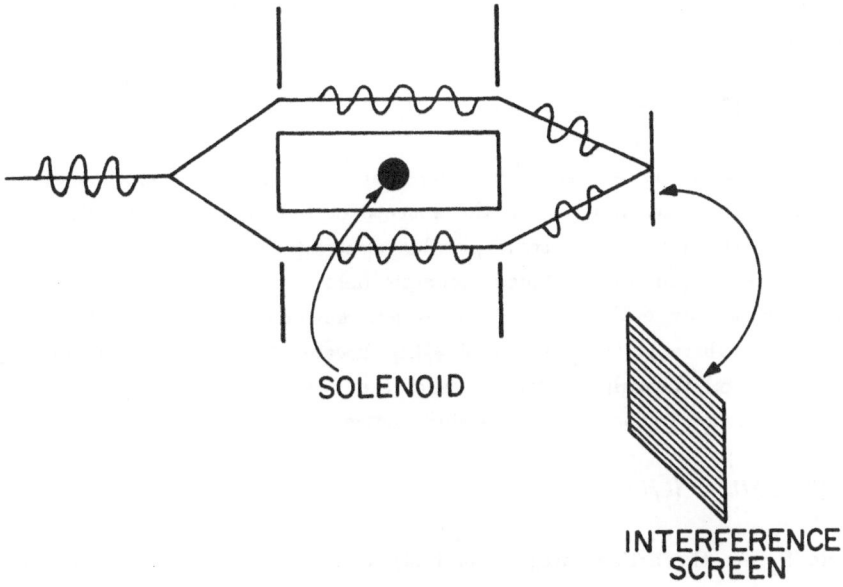

Figure 1a

Magnetic AB Effect. The axis of the solenoid is perpendicular to the page. The wave function is a split plane wave.

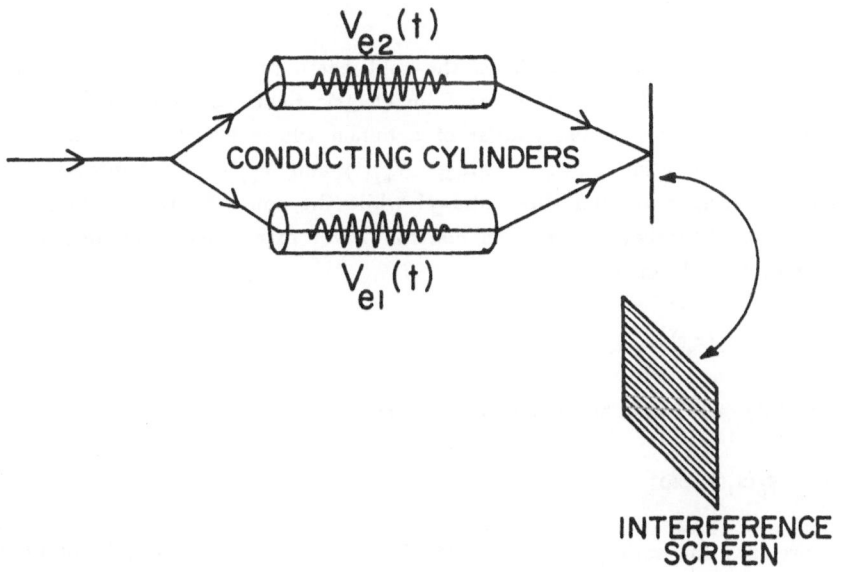

Figure 1b

Electric AB effect. $V_{e1}=V_{e2}=0$ except when the wave packet is shielded from the electric field.

where $A_e(x)$ is the vector potential due to the excluded magnetic field and the $S(x)$ is the line integral

$$S(x) = -(e/c)\int^x A_e(x')\cdot dx' \ , \tag{2.3}$$

and the path of integration is taken along the arm of the interferometer containing the point x. $\psi_0(x)$ is the wave function in the absence of the excluded magnetic field represented by $A_e(x)$, and V_0 represents possible electrostatic potentials to steer the beam which do not depend upon the excluded magnetic field.

If the magnetic flux Φ through the coil is nonvanishing, the vector potential $A_e(x)$ cannot vanish everywhere in the support of $\psi_0(x)$, because $\int A_e(x)\cdot dx$ on a closed path drawn around the coil through the two arms of the interferometer is equal to Φ.

In the interference region, the phase shift between the two beams is

$$\Delta\phi = (S_2 - S_1)/\hbar = (e/\hbar c)\Phi \ , \tag{2.4}$$

where S_2 and S_1 are the action integrals of (2.3), calculated along the upper and lower arms of the interferometer.

The phase shift $\Delta\phi$ between the beams in the two arms of the interferometer is gauge invariant, as it must be, depending only upon the magnetic flux through the excluded region. The interference pattern is therefore a periodic function of that magnetic flux, with period equal to London's unit,

$$\Phi_0 = 2\pi\hbar c/e \ . \tag{2.5}$$

In the electric version of AB effect, the split beam progresses through ideal conducting pipes that shield the electrons from electric fields as shown in Fig. 1b. In this case, the incident beam must consist of a bunch whose length is much smaller than the length of the conducting pipes. Voltages $V_{e1}(t)$ and $V_{e2}(t)$ are impressed on the two pipes, but only during a limited time interval while the split electron beam is deep inside one pipe or the other, so that an electron never experiences any local electric field. Now the Hamiltonian is given by

$$H = H_0 - eV_e(x,t) \ , \tag{2.6}$$

where $H_0 = -(\hbar^2/2m)\nabla^2$ and the wave function is

$$\psi(x,t) = \psi_0(x,t) \ \exp\{-iS_{e\ell}(x,t)/\hbar\} \ , \tag{2.7}$$

where ψ_0 represents the split wave packet in the absence of the external potential $V_e(x,t)$ and

$$S_{e\ell}(x,t) = - e\int_0^t V_e(x,t')dt' \ . \tag{2.8}$$

When the two packets reach the point x in the interference region at some time t after $V_e(x,t)$ has returned to zero everywhere, their relative phase is shifted by the amount

$$\Delta\phi = [S_{2e\ell}(x,t)-S_{1e\ell}(x,t)]/\hbar \; , \qquad (2.9)$$

and that shows up as an observable change in the interference pattern that depends upon the potentials impressed on the two pipes at earlier times t' when the electrons were inside the pipes and experienced no local electric field.

Equation (2.7) gives a solution of the Schroedinger equation

$$i\hbar(\partial\psi/\partial t) = [H_0-eV_e(x,t)]\psi \qquad (2.10)$$

although $\nabla V(x,t)$ vanishes wherever $\psi_0(x,t)$ does not vanish. That is mathematically the essence of the electric AB effect. To achieve that and still get a phase shift between the two beams, we need a region between them where the wave function $\psi_0(x,t)$ vanishes and the electric field, $-\nabla V$, does not vanish. The electron must therefore be confined to a multiply connected region surrounding the excluded electric field, but now that is a space-time region and the periodicity in the external field involves a space-time integral. The electric AB effect will be pursued further in Appendix B.

3. CLASSICAL THEORY

There is no Aharonov-Bohm effect in classical physics. AB effect enters quantum mechanics through the appearance of the electromagnetic potentials V_e and A_e in the Hamiltonian and consequently in the Schroedinger equation. The local Maxwell fields E and B entered Sect. 2 only in the discussion, never in the equations of motion.

When classical theory is presented in the Lagrangean or Hamiltonian formulation, the potentials appear just as they do in quantum theory. However, we know that those formulations of classical physics are equivalent to Newton's laws, so the motion of a charged particle is completely determined by the local electric and magnetic fields that act upon it. Newton's second law and the Lorentz force equation give

$$m(d^2r/dt^2) = -e[E+(v/c)\times B] \; , \qquad (3.1)$$

and nothing more is needed. To remove this feature of the classical theory in the case of a multiply connected region is not a promising enterprise because the local conservation of energy and momentum between the particles and fields depends upon it. Therefore, it is no surprise that the AB effect depends upon the flux or the action in units proportional to Planck's constant $\hbar$, which is peculiar to quantum theory.

Attempts have nevertheless been made to obtain AB effect from classical or semiclassical theory by invoking a reaction on the beam particle which results from its action on the source of the excluded external field [14,15]. That too is an unpromising way to try to explain an interference pattern or a scattering cross section, because for small e the amplitudes would be proportional to e^2 and cross sections to e^4, while quantum mechanics finds them proportional to e and e^2 respectively.

The main point of the attempt based on a classical reaction force appears most simply in the magnetic AB case. The essence of the argument is that the energy has the form

$$E = (1/8\pi)\int[B_e^2+2B_e\cdot B_p+B_p^2]d^3x + (1/2)mv^2 , \qquad (3.2)$$

where B_e is the fixed external field due, for instance, to a current in a solenoid, and B_p is the magnetic field due to the motion of the charged beam particle. There may be additional terms involving the source of the current, but they don't change the argument. Since the B_e^2 term is fixed, the sum of the other terms should be constant. For constant particle velocity v, both the kinetic energy term and the B_p^2 term would be constant. However, the $B_e\cdot B_p$ term is certainly not constant as the particle approaches and then recedes from the solenoid. Therefore the velocity cannot be constant.

That argument is incorrect. In subtracting the external field energy $(1/8\pi)\int B_e^2 d^3x$ from both sides of (3.2), one is subtracting infinite quantities which differ by a finite, time-dependent amount. More careful analysis shows that the finite error precisely cancels the variation in the $B_e\cdot B_p$ energy [4,5]. Then the sum of the kinetic energy plus the B_p^2 energy remains constant, as expected for constant velocity.

The point being made here is analogous to the familiar elementary energy analysis of an elastic collision between Jupiter and a meteor. The meteor is accelerated, but Jupiter is undeflected. The increased energy of the meteor is compensated by the finite change in Jupiter's energy. For a head-on collision,

$$\Delta v_J = \frac{2M_m(v_m-v_J)}{M_J+m_m} . \qquad (3.3)$$

In the limit where Jupiter is infinitely massive, $\Delta v_J \to 0$, but the energy shift,

$$\Delta E_J = \frac{2M_m(v_m-v_J)}{[1+(M_m/M_J)^2]} [v_J+ (M_m/M_J)v_m] , \qquad (3.4)$$

remains finite and obeys

$$\Delta E_J = -\Delta E_m + 2M_m v_J(v_m-v_J) . \qquad (3.5)$$

The details of the correct calculation for the AB case are given in Appendix A. There it turns out that the kinetic energy of the beam particle is constant throughout the collision and there is no reaction force, all in the limiting case of an externally fixed magnetic field whose infinite energy plays the same reservoir-like role as does $(1/2)M_J v_J^2$ for infinite M_J in the limiting Jupiter model.

4. QUANTUM THEORY

Here I describe the general basis of the magnetic AB effect. The electric phenomenon, which is experimentally formidable and theoretically less clean than its magnetic counterpart, is treated in Appendix B.

Quantum theory unavoidably relies upon the Hamiltonian or Lagrangean formulation of the dynamics, where the local electromagnetic fields disappear from the equations of motion in favor of the scalar and vector potentials. The classical argument that the equations of motion are equivalent to Newton's second law with the local E and B fields does not apply to quantum mechanics, and remote fields may have observable effects in some cases. For instance, if a magnetic field $B_e(x)$ is confined to the interior of a torus from which the electron is excluded [29], the vector potential $A_e(x)$ cannot vanish throughout the region outside the torus, and it appears in the Schroedinger equation. The vector potential cannot be removed from the domain of the electron by a gauge transformation because

$$\int A_e(x) \cdot dx = \Phi_e \; , \tag{4.1}$$

where the path of integration links the torus and Φ_e is the magnetic flux through the torus.

In the absence of the excluded magnetic field,

$$i\hbar(\partial\psi_0/\partial t) = H_0\psi_0(x,t) = (1/2m)[-i\hbar\nabla+(e/c)A_0(x,t)]^2\psi_0 - eV_0(x,t)\psi_0 \; , \tag{4.2}$$

where $V_0(x,t)$ and $A_0(x,t)$ are the potentials due to ordinary electromagnetic fields that may exist within the domain of the electron. With the addition of an excluded stationary magnetic field whose vector potential is $A_e(x)$,

$$i\hbar(\partial\psi/\partial t) = H\psi(x,t) = (1/2m)[-i\hbar\nabla+(e/c)\{A_0(x,t)+A_e(x)\}]^2\psi - eV_0(x,t)\psi \tag{4.3}$$

Formally, H and H_0 are related by the gauge transformation

$$U(x) = \exp\{-(ie/\hbar c)\int^x A_e(x') \cdot dx'\} \tag{4.4}$$

$$\psi = U\psi_0 \tag{4.5}$$

$$H = UH_0U^{-1} \; . \tag{4.6}$$

It follows that H and H_0 describe the same physics and the excluded magnetic field $B_e(x)$ has no observable influence on the dynamics of the electron, if Eqs. (4.4)-(4.6) apply.

However, for (4.4)-(4.6) to be meaningful and $\psi = U\psi_0$ to be a single-valued solution of the Schroedinger equation (4.3), U must be a single-valued function of x, independent of the path of integration in the exponent in (4.4). When the domain of x is simply connected, it is sufficient for $B_e(x)=\nabla\times A_e(x)$ to vanish everywhere within it. Then $\int^x A_e(x')\cdot dx'$ is independent of the path of integration, $U(x)$ is single valued, and there can be no observable effect of the excluded magnetic field. But when the domain of the electron is multiply connected as in Fig. 2, and the magnetic field is confined to a region whose topology is that of an excluded cylinder or torus, (4.4) shows that $U(x)$ may not be single valued even if $B_e(x)$ vanishes everywhere in the domain of the electron. Then there is no gauge transformation to connect H_0 with H, and an observable AB effect is possible; the motion of the electron may depend upon the magnetic flux Φ_e through the hole in the electron's domain.

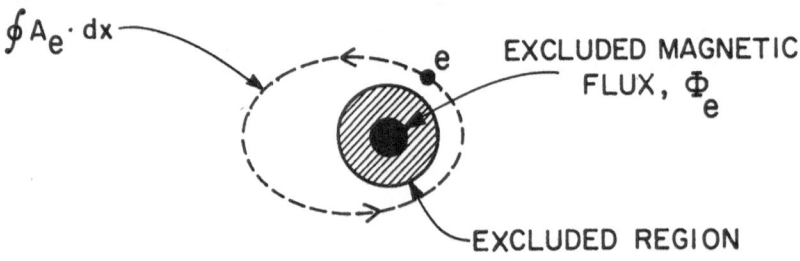

Figure 2

There is an exceptional case. Because only U has to be single valued, not $\int A_e(x)\cdot dx$, the AB effect disappears when the excluded flux $\Phi_e=\oint A_e(x)\cdot dx$ is an integer multiple of Φ_0, i.e. when

$$\Phi_e = n(2\pi\hbar c/e) \ . \tag{4.7}$$

In that case, integrating around the excluded flux changes U by the factor $\exp\{2\pi i\}$, and it remains single valued.

More generally, all observable phenomena depend only upon the flux Φ_e, through the excluded region, and have period Φ_0.

5. BOUND STATE AHARONOV-BOHM EFFECT

The simplest exactly solvable example of AB effect exhibits all the general features of the bound state problem. Consider an electron constrained to move on the circumference of a circle of radius r in the xy plane, as in Fig. 3. An external magnetic flux Φ goes up the z axis and returns uniformly along the surface of a cylinder whose radius is greater than r, so that there is no magnetic field at radius r where the electron moves.

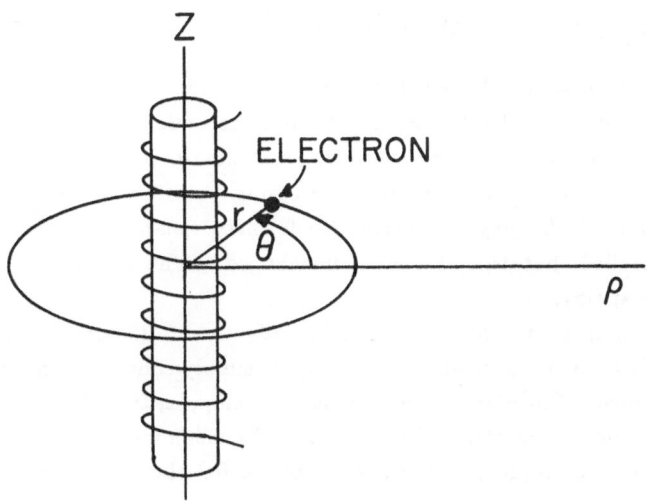

Figure 3

In the gauge where $\nabla \cdot A$ vanishes,

$$A_\theta = \Phi/2\pi r$$

$$A_\rho = A_z = 0 \ . \tag{5.1}$$

The Hamiltonian for an electron of mass m is

$$H = (1/2mr^2)[L_z + r(e/c)A_\theta]^2$$

$$= (1/2mr^2)[L_z + e\Phi/2\pi c]^2 \ . \tag{5.2}$$

The bound-state wave functions and energies are

$$\psi_\ell(\theta) = (2\pi)^{-1/2} \exp\{i\ell\theta\} \tag{5.3}$$

$$E_\ell = (1/2mr^2)[\ell\hbar + e\Phi/2\pi c]^2$$

$$= (\hbar^2/2mr^2)(\ell + \Phi/\Phi_0)^2 \ , \tag{5.4}$$

where ℓ are the integers. The state ψ_ℓ has definite canonical angular momentum L_z and kinetic angular momentum K_z, given by

$$L_z = \ell\hbar \tag{5.5}$$

$$K_z = mr^2\dot{\theta} = (L_z + e\Phi/2\pi c) = \hbar[(\ell + (\Phi/\Phi_0)] \ , \tag{5.6}$$

and the Hamiltonian is equal to the kinetic energy $K_z^2/2mr^2$.

Equations (5.4) and (5.6) clearly display the flux dependence of the energy spectrum and kinetic angular momentum, both measurable quantities in principle. Both spectra are periodic in Φ with period Φ_0, as expected. Increasing the flux by one unit of Φ_0 causes $\psi_{\ell+1}$ to take on all the physical properties formerly possessed by ψ_ℓ. Thus the states have been relabelled, but the physics is unchanged. The corresponding gauge transformation is $U=\exp\{i\theta\}$.

The assumption that ℓ are the integers independently of the flux is the assumption that the wave functions must be single valued. In a simply-connected space that added assumption is not needed; non-integer ℓ would lead to singularities that would spoil the solution of the Schroedinger equation. The reasons why abandoning that assumption in the multiply-connected case would make grave problems for the theory will be discussed in Sects. 6 and 9.

Making the motion three dimensional by allowing the electron to move in a torus instead of on a circle changes nothing important. Then we have

$$H = (1/2m)(p_\rho^2 + p_z^2) + (1/2m\rho^2)(L_z + e\Phi/2\pi c)^2 \tag{5.7}$$

$$\psi_{kn\ell}(z,\rho,\theta) = \chi_{kn\ell}(z,\rho) \ \exp\{i\ell\theta\} \tag{5.8}$$

$$(1/2m)(p_\rho^2 + p_z^2)\chi_{kn\ell} + (\hbar^2/2m\rho^2)[\ell + (\Phi/\Phi_0)]^2\chi_{kn\ell} = E_{kn\ell}\chi_{kn\ell} \ , \tag{5.9}$$

where k and n are the z and ρ quantum numbers. The energy eigenvalues $E_{kn\ell}$ depend upon the flux through the latter's influence on the centrifugal barrier height parameter,

$$Q = K_z^2/2m = (\hbar^2/2m)[\ell + (\Phi/\Phi_0)]^2 \ . \tag{5.10}$$

For example, if the cross section of the torus is small compared to its radius r, the low-lying states $kn\ell$ all involve only the ground state of the ρz motion, and the spectrum is the same as that for the motion on a circle, except for an additive constant. In the general case, the energy spectrum always depends upon the excluded flux because

$$E_{kn\ell} = <kn\ell|(1/2m)(p_\rho^2 + p_z^2) + (Q/\rho^2)|kn\ell> \tag{5.11}$$

$$\frac{dE_{kn\ell}}{d\Phi} = \frac{\hbar^2}{m\Phi_0}\left(\ell + \frac{\Phi}{\Phi_0}\right)\frac{dE_{kn\ell}}{dQ} \tag{5.12}$$

and the right-hand side of (5.12) cannot vanish when $\Phi \neq -\ell\Phi_0$ because

$$\frac{dE_{kn\ell}}{dQ} = <kn\ell|(1/\rho^2)|kn\ell> \neq 0 \ . \tag{5.13}$$

6. THE CENTRAL ROLE OF QUANTIZED ANGULAR MOMENTUM

The quantization of the canonical angular momentum in discrete eigenvalues independent of the magnetic flux plays a central role in the bound state theory by implying quantization of the centrifugal barrier height parameter Q, with flux-dependent eigenvalues [4,30]. Since the energy eigenvalues depend upon the barrier height, the spectrum must depend upon the flux. An analogous consequence may be anticipated for scattering theory. An inverse square law potential scatters particles, so the scattering must depend upon the eigenvalues of Q and therefore upon the magnetic flux in the excluded region. In classical theory, where all values of the canonical and kinetic angular momentum are allowed, no such quantization of Q arises to demand the existence of the AB effect.

Must the eigenvalues of the canonical angular momentum be independent of the flux when the electron moves in a multiply-connected region? The conventional answer is yes. One can imagine turning on or off a cylindrically symmetric magnetic field in the excluded region. The Hamiltonian for the motion of the electron will be time dependent, but it will nevertheless commute with the electron's canonical angular momentum operator, L_z. Then the canonical angular momentum will be a constant of the motion and its eigenvalues will have to be flux independent.

That argument tacitly assumes that the canonical angular momentum operator L_z is to be identified with the generator of rotations around the z axis. The rotation generator is surely conserved under the assumption of cylindrical symmetry, and that justifies the constancy of L_z. In truth, however, symmetry alone does not necessarily exclude the possibility that L_z differs from the rotation generator by a flux-dependent constant, and it is safer to consider the dynamics.

To model the turning on of the flux, consider an infinite cylinder of radius a, situated along the z axis, and suppose a uniform surface current is turned on suddenly at time t=0, flowing in the θ direction. The current I(t) per unit length of cylinder is given by

$$I(t) = \begin{cases} 0 \text{ for } t < 0 \\ I_0 \text{ for } t \geq 0 \end{cases} . \tag{6.1}$$

The vector potential can be obtained from the retarded solution of Maxwell's equations. In the symmetric gauge, where $\nabla \cdot A = 0$ and $V=0$, the only non-vanishing component is $A_\theta(\rho,t)$.

For convenience, a scaled vector potential $g(\rho,t)$ is defined by

$$A_\theta(\rho,t) = (\Phi_\infty/2\pi\rho) \, g(\rho,t) , \tag{6.2}$$

where $\Phi_\infty = 4\pi^2 a^2 I_0/c$. The scaled vector potential $g(\rho,t)$ is calculated for $\rho > a$ in Appendix C. Figure 4 exhibits its qualitative features.

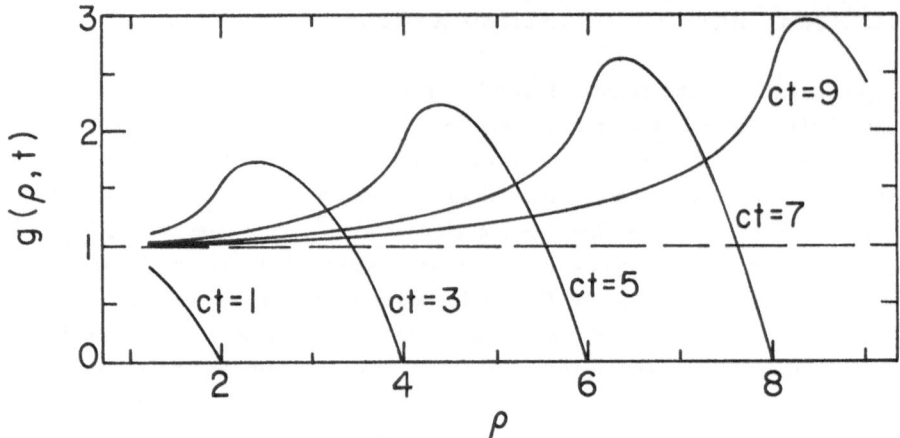

Figure 4

In particular, g vanishes for ct<ρ-a, and g approaches unity for ct>>ρ+a. Thus, $A_\theta(\rho,t)$ is an expanding wave whose front moves out from ρ=a with the velocity of light. At any ρ, the wave front arrives at time t=(ρ-a)/c. For a time interval lasting several times (ρ/c), while A_θ rises and then falls, there is a local field in the θ direction and a magnetic field in the z direction. Then the local electric and magnetic fields die out, and (6.2) shows that A_θ approaches an asymptotic value which corresponds to a steady current I_0.

Now consider the motion of a classical electron with position r(t), under the influence of this expanding wave and possibly some additional forces that have no θ component. The rate of change of K=(r×mv) equals the torque due to the expanding electromagnetic field.

$$dK/dt \;=\; r^{\times}(-e)[E(r)+(v/c)^{\times}B(r)] \tag{6.3}$$

$$dK_z/dt \;=\; (e/c)\rho[(\partial/\partial t)A_\theta(\rho,t)+(\dot{\rho}/\rho)(\partial/\partial\rho)\rho A_\theta(\rho,t)]$$

$$=\; (e/c)(d/dt)\rho A_\theta \;=\; (e\Phi/2\pi c)(dg/dt) \tag{6.4}$$

Since g(ρ,0)=0 and g(ρ,∞)=1 for all ρ, the total change in K_z from t=0 until the time when all fields have died out in the region where the electron moves is given by

$$\Delta K_z \;=\; e\Phi_\infty/2\pi c \;. \tag{6.5}$$

This result is exact in quantum mechanics as well as in classical dynamics because the shift in the kinetic angular momentum is a c-number, commuting with all dynamical variables.

Equation (6.5) explains the non-integral values of the kinetic angular momentum in AB effect with a stationary magnetic field as the result of the torque exerted on the electron by the local E and B fields at past times when the flux was being turned on or when the electron transversed the return flux region. In principle, the time integral of that torque is independent of the electron's motion and of the electron's distance from the excluded flux region at the turning-on time.

Equations (5.6) and (6.5) together give a dynamical justification for the conservation of the canonical angular momentum L_z (in the gauge where $\nabla \cdot A$ vanishes) as the excluded flux varies. The same physics can be expressed in somewhat different form by observing that the total angular momentum J_z in the the coupled system, electron plus external magnetic flux, is given by

$$J = r \times mv + (1/8\pi c)\int x \times [E(r,x) \times B_e(x)]d^3x \; , \tag{6.6}$$

where $B_e(x)$ is the external magnetic field and

$$E(r,x) = -e(x-r)/|x-r|^3 \tag{6.7}$$

is the electric field at x whose source is the electron at r. The second term in (6.6) may be integrated to give

$$J_z = \{r \times [mv - (e/c)A_e(r)]\}_z = L_z \; . \tag{6.8}$$

Thus, L_z is equal to the total angular momentum, kinetic plus electromagnetic, which should be conserved as the flux is turned on.

This derivation is not formally gauge invariant and cannot be, because the total angular momentum is gauge invariant and the canonical angular momentum is not. The gauge where $\nabla \cdot A_e$ vanishes is a good one for present purposes, because that is the unique gauge in which the vector potential exhibits the symmetry of the electromagnetic field. All the same physical results can be obtained more awkwardly in a general gauge. The spectrum of L_z remains gauge invariant, as it must.

To evaluate the integral in (6.6), it is necessary to include the return flux at infinity or at some large distance. If that is not done, L_z continues to be conserved but the total angular momentum J_z is not; some angular momentum is carried off to infinity by the expanding electromagnetic wave when the flux is turned on, and then not included in (6.6). However, all the measurable phenomena are unchanged because L_z continues to be conserved and quantized in integers.

Additional details of the angular momentum analysis have been given elsewhere [30].

7. SCATTERING STATE AB EFFECT

The prototype for scattering and diffraction problems is the scattering of a beam of electrons moving in the -x direction, by a flux line along the z axis, with zero radius and finite flux Φ. I will not repeat the details of that calculation, which were given by

Aharonov and Bohm [2]. However, I will review some unusual features of their scattering wave function which bear upon the existence and the meaning of AB effect in standard quantum mechanics.

The idealized scattering experiment is illustrated in Fig. 5, with the magnetic flux separated from the electron beam by a cylinder of finite radius a. The calculation of Ref. [2] corresponds to the case a=0. The incident beam region initially contains a Gaussian wave packet centered on the wave function represented in (7.3) below.

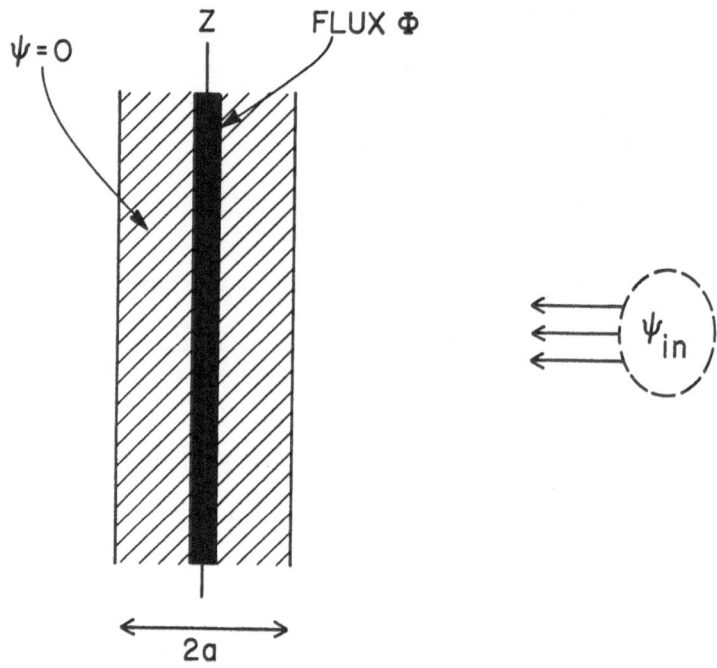

Figure 5

For a flux line of vanishing radius,

$$-i\hbar \frac{\partial \psi}{\partial t} = - \left[\frac{\hbar^2}{2m}\right] \left[\frac{\partial^2}{\partial t^2} + \frac{1}{\rho}\frac{\partial}{\partial \rho}\left(\rho\frac{\partial}{\partial \rho}\right) + \frac{1}{\rho^2}\left(\frac{\partial}{\partial \theta} + i\alpha\right)^2\right]\psi \qquad (7.1)$$

except at the z axis, where $\psi(\rho,\theta,z)$ obeys

$$\psi(0,\theta,z) = 0. \qquad (7.2)$$

Here, as in Ref. [2], the incident wave is taken to enter from the +x direction and move toward the -x direction, so the time-independent wave function obeys

$$\psi = \exp\{-ia\theta\} \exp\{-ikx\} + \frac{f(\theta)}{\sqrt{\rho}} \exp\{ik\rho\} + 0(1/\sqrt{\rho}^3) \qquad (7.3)$$

in the asymptotic region where $\rho \to \infty$. The azimuthal angle θ is defined by the range $-\pi \leq \theta \leq \pi$, and the flux parameter a is given by

$$a = e\Phi/2\pi\hbar c = \Phi/\Phi_0 . \qquad (7.4)$$

The first term in (7.3) is multiple-valued in the sense that the function or its derivative is discontinuous along the negative x axis. The exact ψ contains a compensating term and the complete wave function is single valued everywhere. An example of this compensation is given below in the simple case $a=1/2$.

In the incoming beam region, $x \to +\infty$, the first term in (7.3) obeys

$$mv_x\psi = -\hbar k\psi \qquad (7.5)$$

$$mv_y\psi = mv_z\psi = 0 , \qquad (7.6)$$

where the velocity operator v is given by

$$mv = (\hbar/i)\nabla + (e/c)A . \qquad (7.7)$$

The scattering cross section per unit length of the flux line can be obtained from the outbound flux at $\rho = \infty$, with the usual result,

$$d\sigma/d\theta = (1/k)|f(\theta)|^2 . \qquad (7.8)$$

The unusual first term in (7.3) replaces the incident plane wave of problems without the long range vector potential due to the flux line. If the factor $\exp\{-ia\theta\}$ were omitted, v_y would be proportional to $(1/\rho)$ and a Gaussian asymptotic wave packet made of such plane waves would miss its target at the origin by an infinite amount.

For half-integer values of the flux parameter a, Aharonov and Bohm [2] found an exact solution of the Schroedinger equation:

$$\psi = \exp\{-ia\theta - ikx\} \int_0^{\sqrt{k(r+x)}} \exp\{it^2\}dt . \qquad (7.9)$$

This agrees with (7.3) asymptotically and has the merit of simplicity. The wave function is manifestly single valued along the negative x axis for half-integer a, because the integral vanishes there and is an even function of y, as needed to remove the multiple-valuedness of the factor $\exp\{-ia\theta\}$. From the asymptotic limit of (7.9) for large ρ and fixed θ, one finds

$$f(\theta) = \frac{\exp\{-i\theta/2\}}{(2\pi)^{1/2} \cos(\theta/2)} \qquad (7.10)$$

$$d\sigma/d\theta = (1/2\pi k) \cos^{-2}(\theta/2) \ . \tag{7.11}$$

Ref. [2] also solved the scattering problem for general α, finding a more complicated and less transparently single-valued wave function, and a cross section equal to that in (7.11), multiplied by $\sin^2(\pi\alpha)$.

The interpretation of these results as illustrating a flux-dependent scattering by an excluded magnetic field is partially obscured by the use of a zero-radius flux line with no return flux. One may question in what sense the magnetic field is separated from the domain of the electron when a=0, even though it turns out that all eigenfunctions of the Hamiltonian vanish at ρ=0. Having no return flux is possibly unphysical although there is nothing in Maxwell's equations against it if one does not object to an infinite solenoid, and it is precisely this feature which results in the distorted incoming wave at infinity in place of the familiar plane wave. However, it was shown by Peshkin, Talmi, and Tassie [4] and by Tassie [29], that those objections are irrelevant. Surrounding the flux line by an excluded cylinder of finite radius a, and confining the magnetic flux to e.g. radius a/2, as in Fig. 5, still leads to a flux-dependent cross section and the wave function goes smoothly over to that of Aharonov and Bohm in the limit a→0. Introducing a second cylinder to contain the return flux still leads to a solvable Schroedinger equation in the special case where α is a half-integer, and the Aharonov-Bohm effect does not go away; the cross section demonstrably depends upon the excluded flux, and the incoming wave function approaches a plane wave at infinity. The currents at infinity can be avoided by substituting a finite toroidal solenoid for the infinite cylinder [29].

8. LOCALITY AND CAUSALITY

The AB effect does not violate Einstein causality; it sends no signal faster than light. To investigate that possibility, consider a diffraction experiment in which a beam of electrons is split by an impenetrable solenoid at a time before impinging upon a detector screen, as in Fig. 6.

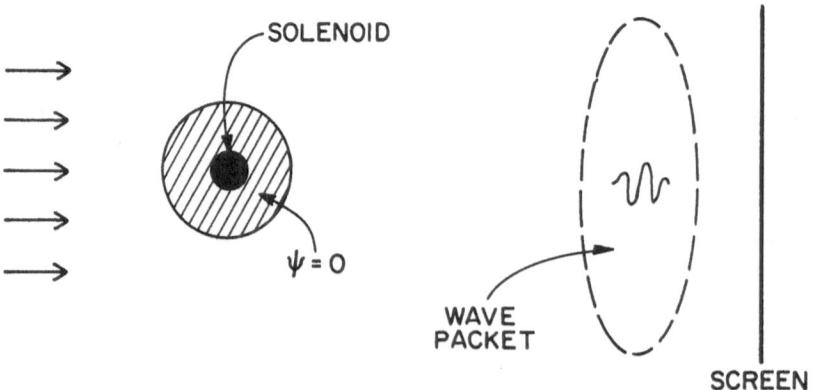

Figure 6

Some current is suddenly turned on at the solenoid when the electron's split wave packet has passed the cylinder and is just about to strike the screen. The vector potential at the position of the electron depends upon the contemporaneous flux through the solenoid.

$$A_\theta(\rho,t) = \Phi(t)/2\pi\rho .$$ (8.1)

Since the vector potential does influence the time development of the wave function, and through it the observable diffraction pattern, one may ask whether the changed diffraction pattern receives a superluminal signal.

The answer to that question is no. The Maxwell equations forbid arbitrary $\Phi(t)$, and they forbid vanishing electromagnetic fields outside a cylinder containing a time dependent magnetic flux. Even if the current in an infinite solenoid is turned on suddenly, as in Sect. 6, the resulting vector potential moves out with velocity c, as illustrated in Fig. 4. Then if

$$H'(r,t) = -(e/c)v \bullet A(r,t) + (e^2/2mc^2)A^2$$ (8.2)

represents the interaction with that vector potential,

$$\psi(r,t) = \psi_0(r,t) + \int \psi(r',t') \, H'(r',t') \, G_0(r',t',r,t) \, d^3r'dt' ,$$ (8.3)

where ψ_0 and G_0 are the wave function and Green's function in the absence of the vector potential H'. In (8.3), $H'(r',t')$ carries no information faster than light because it depends only upon the causal vector potential $A(r',t')$ of Sect. 6. The wave function $\psi(r,t)$ nevertheless depends upon $\Phi(t)$ in the past light cone if G_0 is the Green's function for the nonrelativistic Schroedinger equation, but that is a fault of the nonrelativistic theory and has nothing to do with the electromagnetic interaction. The Green's function for any relativistic theory vanishes for $|r'-r| < c|t'-t|$, and then (8.3) confirms that the AB effect does not contradict causality.

Locality has several different aspects. Firstly, one may inquire about Ehrenfest's theorem. In the AB effect, there is no electromagnetic force on the electron, yet the electron is scattered in a flux-dependent way and the momentum transfer to the electron in the scattering process depends upon the flux. In the absence of boundaries where the wave function is made to vanish, that would contradict the Ehrenfest theorem, which requires that the expected rate of momentum transfer to the electron, $(d/dt)<\psi|mv|\psi>$, should equal the expectation of the electromagnetic force, $(i/\hbar)<\psi|[H,mv]|\psi>$. However, in the multiply-connected region required for the AB effect, there is also a surface contribution which represents the momentum transfer upon reflection from a barrier, and

$$(d/dt)<\psi|mv|\psi> = (i/\hbar)<\psi|[H,mv]|\psi> + \int[(mv\psi)(d\sigma \bullet \nabla\psi^*)-\psi^*(d\sigma \bullet \nabla)mv\psi]$$ (8.4)

For the AB effect, the first term in the r.h.s. of (8.4) vanishes. The second term, which represents the force experienced by the electron upon reflection from the barrier, can be evaluated to show that the force expressed in (8.4) agrees with the rate of momentum transfer calculated from the cross section, $\int (d\sigma/d\theta)\hbar q(\theta)d\theta$, where $q(\theta)=k(1+\cos\theta)$. The details of that calculation are given in Appendix D for the special case where the exact wave function was calculated by Aharonov and Bohm.

The consistency of the cross section with the momentum transfer at the barrier shows that the electron strikes the barrier in a flux dependent way and requires no nonlocal electromagnetic force to experience a flux dependent momentum transfer.

That result is not paradoxical because it is impossible to assemble an incident wave packet in the incoming region of Fig. 5 with objective properties that are independent of the flux. Nevertheless, one may ask whether the manifestation of nonlocality in the electromagnetic interaction has simply been moved from the scattering center to the incident beam region. In one sense, the answer to that question is also no. An incident wave packet in the presence of nonvanishing flux contains partial waves whose kinetic angular momentum spectrum, in principal a measurable property,[*] differs from the integers by Φ/Φ_0. In Sect. 6 it was seen that this shift in kinetic angular momentum resulted from a local electric field at the position of the electron at some past time, and delivered an angular impulse independent of the electron's motion and distance from the flux bearing cylinder.

In another sense, this does appear to demonstrate a nonlocality in the theory. The electron could have been created as one member of a pair after the magnetic flux in the cylinder was established and the expanding wave front had long passed the incident beam region. Then that electron never experienced any local electromagnetic field, but the kinetic angular momentum values with which it was allowed to be created, and therefore its future scattering, nevertheless depend upon the remote magnetic flux. The available Hilbert space was changed by the passing local field, independently of the presence or absence of the electron at the time when the field was present.

However, this kind of nonlocality appears to be a feature of quantum mechanics which is not restricted to the AB effect or to the electromagnetic interaction, or even to multiply connected regions. Consider an alpha particle created as one member of a pair in the incident beam region, long after a target alpha particle has been constrained to the origin. Because alpha particles are bosons, the incident wave packet contains only even angular momentum values, and that has measurable consequences for elastic scattering and for reactions such as $a+a \rightarrow {}^6\text{Li}+\text{d}$. Thus the alpha particle at the origin has had an objective nonlocal influence in the incident beam region, albeit one that cannot be measured in the incident beam region.

---

[*]This property cannot be observed in any local measurement in the incident beam region. An observation with $\Delta L_z < \hbar$ requires $\Delta\theta > 2\pi$. Thus the uncertainty principle protects this aspect of the theory from experimental challenge, in the usual way.

9. LESSONS FROM THE EXPERIMENT

Experimentally, it is now clear that the effects of magnetic fields in inaccessible regions are correctly described by ordinary quantum mechanics. We are instructed to use the single valued solution of the Schroedinger equation, at least in those cases where the experiment has been done. What do we learn from that?

Firstly, the underlying principle of gauge field theory is tested in a uniquely direct experiment, and confirmed. The local Maxwell fields alone are insufficient to describe electromagnetism. The local four-vector potential is clearly more than sufficient, because different but gauge-equivalent potentials describe the same physics. The quantities which are necessary and sufficient to represent the effects of a magnetic field are the phase factors

$$U = \exp\{(ie/\hbar c)\int A(r) \cdot dr\} \tag{9.1}$$

for all paths of integration within the domain of the electrons. No previous experiment demonstrated this unambiguously, because U depends upon $B(r)$ in a simply-connected domain, even though U is expressed as a function of A. Only the AB effect produces non-trivial U with vanishing $B(r)$ over the entire domain of r. The importance of this test has been emphasized by C. N. Yang for many years [7,8].

A similar statement applies to the electric AB effect, where

$$U = \exp\{(ie/\hbar)\int V(t)dt\} \ . \tag{9.2}$$

This feature of the theory has not been tested by experiment, but it can in principle be obtained from the magnetic case by a kind of Lorentz symmetry argument (Appendix B).

From the perspective of quantum mechanics in an external magnetic field, the issues of:

1)   what boundary condition to use in a multiply connected region, and
2)   how those boundary conditions depend upon the excluded flux

are actually distinct questions, both addressed by the high precision experiments of Tonomura. All the arguments reviewed in this paper and in most papers to show that the theory would encounter contradictions without the AB effect really only deal with the effects of changing the magnetic field, item 2 above [31]. If one sticks to single-valued wave functions with all magnetic fields, the theory is consistent with the conservation of angular momentum, and with the experiments. The shift in the kinetic angular momentum eigenvalues agrees with the integrated torque on the electron during the period when the magnetic field is changed, no matter how far away the electron may be and no matter how it moves. An equal and opposite shift of the angular momentum in the crossed fields, E due to the electron, and B external, compensates for the altered kinetic angular momentum of the electron so that the total angular momentum is conserved. If the boundary condition is made flux dependent to eliminate the AB effect, angular momentum will not be conserved and it is hard to see how the dynamics of the electromagnetic field can be modified to deal with that. In fact, one can make similar statements about the Zeeman effect on the orbital motion in an atom, but the AB effect is a much clearer case because no forces are exerted by the magnetic field. Thus, the recent experiments confirm these fundamental aspects of the theory in a new regime where the issues are clearest.

However, the angular momentum analysis does not actually require the use of single-valued wave functions in a multiply-connected region. It is enough that the boundary condition should not change with the flux so that the total angular momentum can be conserved when the flux is changed. In other language, multiple connectivity introduces inequivalent representations of the commutation relations, with inequivalent physical consequences. This has been known since the early work of Weyl, and its consequences for the AB effect have also been discussed [32]. A similar reservation exists with respect to gauge theory, which has not been proved to exclude the possibility of inequivalent representations in a multiply connected region. The question remains an experimental one even if one accepts all the conventional theoretical wisdom. It is answered convincingly by the Tonomura experiments, where the interference patterns inside and outside his rings line up precisely when the magnetic flux either vanishes or equals an integer times Φ_0.

The physics at issue in the inequivalent representations is exhibited fully by considering the motion of a particle outside of an excluded cylinder[*] and using the commutation relations appropriate to rotations in two dimensions. In this approach there is no need to argue about multiple-valued wave functions.

$$[L_z, \theta] = -i\hbar .$$ (9.3)

(Strictly, there is no variable θ because of the 2π ambiguity, and I should use $[L_z, u] = \hbar u$, where $u = \exp\{i\theta\}$. That turns out not to matter for present purposes.)

The eigenvalues $\ell\hbar$ of L_z have the spectrum

$$\ell = \beta + \mu ,$$ (9.4)

where μ are the integers and β is an arbitrary constant in $0 \leq \beta < 1$.

Equation (9.4) can be proved by considering a 2π rotation around the z axis, $R(2\pi)$, which must carry every state

$$\psi = a_1|\ell_1\rangle + a_2|\ell_2\rangle$$ (9.5)

into itself times a possible phase factor.

---

[*]It is simplest to consider the case of cylindrical symmetry and make use of the angular momentum, but cylindrical symmetry is not essential to the physical results. The excluded region could equally well be the finite toroidal region of the Tonomura experiments.

$$R(2\pi)\psi \;=\; \exp\{i2\pi\ell_1\}\; a_1|\ell_1> \;+\; \exp\{i2\pi\ell_2\}\; a_2|\ell_2> \tag{9.6}$$

For $R(2\pi)\psi$ to equal ψ times a phase factor for all ψ, the eigenvalues ℓ must be spaced by integers, and that implies (9.4). There is nothing in the principles of quantum mechanics to require $\beta=0$ in the general case, where regularity at the z axis is not needed because of the excluding cylinder.

Now the centrifugal potential is $(\hbar^2/2\mu\rho^2)(\mu+\beta)^2$, which depends upon β. Different β lead to different energy spectra for bound states and different cross sections for scattering states, all in the absence of any magnetic field.

In the language of wave functions, one can use the multiple valued boundary condition

$$\psi(z,\rho,2\pi) \;=\; \exp\{i2\pi\beta\}\psi(z,\rho,0) \;, \tag{9.7}$$

realizing L_z by

$$L_z \;=\; -i\hbar[\partial/\partial\theta-(1-\exp\{i2\pi\beta\})\delta(\theta)] \;, \tag{9.8}$$

with eigenvalues

$$\ell\hbar=\hbar(\mu+\beta) \;. \tag{9.9}$$

In principle, if one has an excluded cylinder or torus, only an experiment can determine β. The electron or other particle could be coupled to β through its charge or through some other, possibly unknown quantum number. Tonomura's electron holography experiments are uniquely sensitive to such a hypothetical coupling, and they apparently now exclude $|\beta|>.01$ for the samples tested.

This result goes beyond the AB effect in the sense that it answers question 1 above, by looking for a new phenomenon that would be available only in a multiply connected region but which, unlike AB effect, would not necessarily require any magnetic flux.

Any coupling which preserves time reversal symmetry, or which preserves symmetry under a 180° rotation about an axis perpendicular to the z axis, can only result in β equal to zero or one-half, because for such a theory the negative of an eigenvalue ℓ is also an eigenvalue. However, that does not reduce the interest in experiments to detect nonvanishing $\beta<<1$, if one is willing to give up time reversal symmetry and the relevant rotation symmetry.

It has sometimes been suggested that neither AB effect nor the zero-field effect can exist in principle, because exclusion of the electron is only approximate. In reality the electron must penetrate the excluded region, though its wave function may become tiny. From that point of view, a multiply-connected region has no physical reality. However, one may suppose that even then an exactly vanishing wave function in the "excluded region" is actually the best starting point for a useful theory. The objection must be compared to a claim that there can be no phase transitions in nature because the partition function is always continuous for a finite number of molecules.

In making models for elementary particle theory, one can go further and realize the excluded region literally, even in principle. If an electron is bound by some scalar force to an excluded cylinder of finite or vanishing radius, one has the theoretical possibility of a bound state version of the β coupling, with the angular momentum and energy spectra of the resulting composite system dependent upon β. Such composites have been hypothesized in a model with an electron bound to a a magnetic flux line [23], but their most general properties appear to be the same as one obtains simply by postulating a representation with nonvanishing β; the postulated magnetic flux is irrelevant. Discussions of these composites based on quantum field theory have generally concluded that β must equal zero or one-half [33], but that is a consequence of the theory's time-reversal symmetry and not necessarily general [34].

Finally, it seems useful to note that experiments on the AB effect have important implications for Dirac's charge quantization condition,

$$eg/c = n\hbar/2 \; , \tag{9.10}$$

where g is the hypothetical magnetic monopole charge and n is an integer. Dirac's original theory required a flux string carrying flux $\Phi_D = 4\pi g$ between the magnetic charge and infinity, and other conventional monopole theories do the same thing in one way or another. The flux in the Dirac string is a multiple of London's Φ_0 if and only if (9.10) is obeyed, and only then is the orientation of the string physically unobservable so that the monopole appears as a particle described by the usual dynamical variables. The scattering of electrons by the flux-bearing Dirac string would be an example of AB effect, at least if the electron beam were kept far enough from the monopole so that the string could be considered infinite in both directions for practical purposes. Failure to observe the AB effect for $\Phi \neq n\Phi_0$ would destroy the reason for quantizing the charge product, eg, in Eq. (9.10). Conversely, failure of the AB effect to vanish exactly when $\Phi = n\Phi_0$ would result in the Dirac string's being physically observable even when (9.10) is obeyed, destroying the basis of the monopole theory. These contradictions could not easily be resolved by modifying the monopole theory because the role of the angular momentum in the monopole theory is essentially the same as in AB effect [34], and changing it would require a drastic revision of our understanding of angular momentum in quantum mechanics.

This work was supported by the U. S. Department of Energy, Nuclear Physics Division, under contract W-31-109-ENG-38.

APPENDIX A - ENERGY IN THE MAGNETIC FIELD

The question has frequently been raised [14,15] whether the Hamiltonian formulation of the theory somehow takes account of a reaction force on the beam particle due to the action of the beam on the sources of the external magnetic field. Specifically, the energy in the total magnetic field, which is proportional to $\int B^2 \, d^3x$, contains the time varying interaction term $\int B_e \cdot B_p d^3x$ between the "fixed" external magnetic field B_e and the changing magnetic field B_p whose source is the beam particle. Do the changes in that interaction energy come from the kinetic energy of the beam particle?

That question can be answered reliably by including the source of the external field in the dynamics and taking the appropriate limit, and the answer to the question is no. The interaction energy comes out of the infinite energy in the "fixed" external field and its sources, and the kinetic energy of the beam particle remains constant. The calculation has been given before, both for the example of a mechanical model [4] and in completely electromagnetic terms [5]. Here I follow the electromagnetic approach, including details that should make the external field limit unambiguous.

Consider a truncated cylinder of radius a and length L>>a, centered at z=0 and situated so that the cylinder axis is the z axis. A uniform surface current I per unit length of cylinder circulates in the θ direction. The physical model for this could be based on a tightly wound coil of resistanceless wire. An otherwise free charged particle passes near but outside the coil, near z=0, so that any local magnetic field it experiences will be of order $(a/L)^2$ in comparison with the field inside the coil. The AB case is the limit L→∞, with finite a. (It can also be obtained more generally with finite L by adding magnetic shields, but that is not done here.)

The conserved total energy in this system is

$$E = E_I + E_i + E_p , \tag{A.1}$$

the sum of the energy E_I associated with the current I on the surface of the cylinder, the energy E_p of the free particle, and the interaction energy E_i. Since the magnetic field inside an infinite cylinder is equal to $4\pi I/c$, the magnetic field energy due to the circulating current I is given by

$$E_I = (1/8\pi)\int B_I(x)^2 d^3x = 2(\pi Ia/c)^2 L[1+0(a/L)] . \tag{A.2}$$

The correction term of order (a/L) takes account of end effects and the energy in the return flux, and it will be omitted hereafter.

The current I may be stabilized by approaching the limit L→∞ or by adding some other machinery whose energy is proportional to I^2. Such machinery could e.g. be a second coil with suitably large inductance, or a flywheel attached to a generator to provide the current I. In any case,

$$E_I = (1/2)WI^2 , \tag{A.3}$$

where W is a constant that includes contributions from (A.2) and from the inductance of any attached coil and the kinetic energy of any flywheel. All we need is that E_I is proportional to I^2 and that the fixed external field limit corresponds to $W \to \infty$.

The free particle energy E_p consists of the kinetic energy $(1/2)mv^2$ plus the energy of the magnetic field $B_p(x,t)$ whose source is the moving charge. Since B_p is everywhere proportional to v,

$$E_p = (1/2)Mv^2 , \tag{A.4}$$

where M is some effective mass which takes account of the field energy.

The interaction energy is given by

$$E_i(t) = (1/4\pi) \int B_I \cdot B_p d^3x = (I/c) \int \Phi_p(z,t)dz , \tag{A.5}$$

where $\Phi_p(z,t)$ is the particle-induced magnetic flux through the coil at height z.

$$\Phi_p(z,t) = \int\int B_{pz}(x,y,z,t)dxdy \tag{A.6}$$

The integral in (A.6) is taken over the interior of the coil, $x^2+y^2 \leq a^2$. End corrections and return flux effects have been suppressed for typographic simplicity; they vanish like $(a/L)^2$ in the limit $L \to \infty$.

The time varying B_p induces an electromotive force $\partial\Phi_p/\partial t$ that does work against the current I, thereby changing the energy $E_I=(1/2)WI^2$.

$$dE_I/dt = I\int[-\partial\Phi_p(z,t)/\partial t]dz = -(dE_i/dt) \tag{A.7}$$

Thus, while E_i changes by a finite amount as the beam particle passes the solenoid, the sum (E_I+E_i) remains constant. Then E_p is constant and the interaction energy varies only at the cost of the energy in the external field and its sources.

For finite W, which implies finite length of the coil, the current I and the field due to the coil vary as the particle goes by.

$$\delta I = \delta E_I/(WI) \tag{A.8}$$

In the limit of infinite L, where W is also infinite, the external magnetic field becomes fixed and its infinite energy varies by a finite amount so that the total energy remains fixed.

All this can be said still more generally at the cost of minor additional complication, but with little profit. The energy can be replaced by the Hamiltonian to show that the equations of motion in an external field are correct in the AB limit, which means that v remains constant. In the semiclassical case, use of the correct total Lagrangean removes the phase change which would come about if only (dE_i/dt) were included, and not (dE_I/dt). The limit $(a/L) \to 0$ can also be avoided by other devices to guarantee zero local magnetic field at the position of the beam particle.

APPENDIX B - ELECTRIC AB EFFECT

The electric AB effect, in contrast to the magnetic phenomenon, has not lent itself to experimental test. Moreover, it appears that no exact theoretical treatment has been given. Here I rely upon approximations that should be adequate for nonrelativistic electrons in an interferometer such as that depicted in Fig. 1b.

Ideally, the excluded electric field $E(x,t)$ should obey

$$\psi(x,t)\ E(x,t)\ =\ 0 \qquad\qquad \text{for all } (x,t) \tag{B.1}$$

$$E(x,t)\ =\ 0 \qquad\qquad \text{for } (x,t)\epsilon D\ , \tag{B.2}$$

where the space-time region D consists of the detection area when the two beams have been reunited and the entrance area before they were separated (Fig. 1b).

In an interferometer experiment, the electron enters the apparatus at time t=0 with a wave packet $\psi(x,0)$. At later times when the potential is turned on, the main wave packet has divided into two pieces, each deep inside one arm of the interferometer. Equation (B.1) is obeyed only in the approximation that the wave function can be replaced by zero outside the two main peaks of the wave packets. (Conceptually, the electron could be confined to two moving boxes, one in each arm.) To have this separation of the wave packet during the experiment, but not at the beginning and the end, requires a multiply-connected spatial geometry, so there can be no electric AB effect in a simply-connected region

Where there are time-dependent electric fields, there are vector potentials and magnetic fields. I neglect those and use

$$H\ =\ H_0\ -\ eV_e(x,t) \tag{B.3}$$

$$V_e(x,t)\ =\ 0 \text{ for } (x,t)\epsilon D \tag{B.4}$$

The gauge transformation analogous to (4.4) is

$$\psi(x,t)\ =\ U(x,t)\psi_0(x,t) \tag{B.5}$$

$$U(x,t)\ =\ \exp\{(-ie/\hbar)\int_0^t V_e(x,t')dt'\} \tag{B.6}$$

Eq. (B.5) gives the solution of the Schroedinger equation

$$i\hbar(\partial\psi/\partial t)\ =\ H\psi(x,t) \tag{B.7}$$

if

$$2\nabla\psi(x,t)\cdot\int_0^t E(x,t')dt'+\psi(x,t)\left[\int_0^t \nabla\cdot E(x,t')dt'+(ie/\hbar)\left(\int_0^t E(x,t')dt'\right)^2\right]\ =\ 0 \tag{B.8}$$

for all (x,t), i.e. if

$$\psi(x,t) \ E(x,t') = 0 \ \text{whenever} \ 0 \leq t' \leq t \leq T, \tag{B.9}$$

where T is a time when the wave packet has arrived at the detector.

If (B.9) is obeyed for all $t' \leq t$, then (B.8) is obeyed, $\psi(x,T)$ is just a number times $\psi_0(x,T)$, and the excluded field has had no effect on the motion of the electron. For the AB effect to be observed, (B.9) must fail for some t'; the electron must traverse some region where the electric field has been. In the example of Fig. 1b, that region lies between the right end of the beam tubes and the detection region.

That (B.9) must fail for a physical effect to exist is consistent with the ordinary ideas about causality. One cannot wait for the electron to pass and only later switch on the field to cause a physical effect. This feature arises more transparently in the treatment of the magnetic AB effect (Sect. 8).

In the two-arm spectrometer, where (B.8) fails, (B.6) does not give the solution of the Schroedinger equation in the detection region, but it is correct up to an earlier time τ when the potentials have been turned off but the wave packets have not yet left the shielded arms. For $t>\tau$, $H=H_0$. Therefore, $\psi(x,t)$ for $t>\tau$ is equal to

$$\psi_0(x,t) \ \exp\{-i(e/\hbar)\int_0^\tau V(x,t')dt'\} \tag{B.10}$$

in each arm, and the reunited beams are shifted in their relative phase by the amount

$$-i(e/\hbar)\int_0^\tau [V_1 t')-V_2(t')]dt' \ . \tag{B.11}$$

Then the interference pattern is periodic in

$$\Phi_{e\ell} = c \int_0^\tau \Delta V(t)dt \ , \tag{B.12}$$

with period $\Phi_0=2\pi\hbar c/e$, just as in the magnetic case.

Aharonov [35] points out that there is a kind of Lorentz symmetry in the periodicity. It can be seen by writing the flux in the form

$$\Phi = \int F^{\mu\nu} \ d\sigma_{\mu\nu} \ , \tag{B.13}$$

where F is the Maxwell field tensor. In the magnetic case, the surface elements $d\sigma_{\mu\nu}$ belong to a space-like surface at constant time whose periphery is an orbit surrounding the excluded field. In the electric case, the surface is time-like.

APPENDIX C - TIME-DEPENDENT FLUX

In Section 6, a surface current

$$j_\theta(\rho,t) \;=\; I_0\delta(\rho\text{-a})1(t) \tag{C.1}$$

circulates on the surface of a cylinder of radius a whose axis is the z axis. In the limit $ct \gg \rho$, the magnetic flux through the cylinder and the vector potential outside the cylinder must approach their steady-state values,

$$\Phi_\infty \;=\; 4\pi^2 a^2 I_0/c \tag{C.2}$$

$$A_\theta(\rho,\infty) \;=\; \Phi_\infty/2\pi\rho \;. \tag{C.3}$$

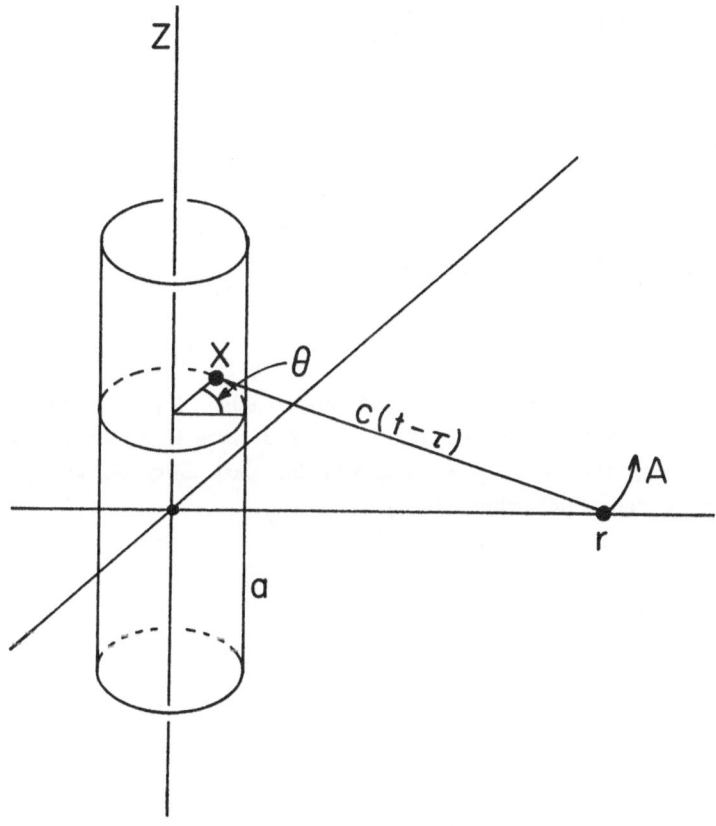

Figure 7

The vector potential A lies in the xy plane.

For finite times, the retarded solution of the wave equation for A is given by

$$A_\theta(\rho,t) = (1/c)\int j_\theta(x,t-\tau)/|x-r|d^3x \ , \tag{C.4}$$

where r and x have cylindrical components given by $r=(0,\rho,0)$, $x=(z,a,\theta)$ as shown in Fig. 7, and the retarded time τ is defined by

$$c^2(t-\tau)^2 = z^2 + \rho^2 + a^2 - 2a\rho\cos\theta \ . \tag{C.5}$$

The z integral can be carried out to give

$$A_\theta(\rho,t) = (\Phi_\infty/2\pi\rho)g(\rho,t) \ , \tag{C.6}$$

where the scaled vector potential $g(\rho,t)$ vanishes for $ct<\rho-a$ and otherwise obeys

$$g(\rho,t) = \frac{2\rho}{\pi a}\int_0^\theta \cos\theta \ \ell n\left[\frac{ct + \sqrt{(ct)^2-\rho^2-a^2+\rho\cos\theta}}{\sqrt{\rho^2+a^2-2a\rho\cos\theta}}\right]d\theta \ . \tag{C.7}$$

The upper limit θ is defined by

$$\cos\theta = \begin{cases} (\rho^2+a^2-c^2t^2)/2a\rho & \text{for } ct<\rho+a \\ -1 & \text{for } ct>\rho+a \ . \end{cases} \tag{C.8}$$

In the limit $(ct-\rho)\to\infty$, (C.7) can be simplified to give

$$g(\rho,\infty) = 1 \ . \tag{C.9}$$

The integral in (C.7) has been calculated numerically. Figure 4 displays the behavior of $g(\rho,t)$ for selected finite values of ct. The wave front moves out from $\rho=a$ with velocity c, passing the radius ρ at $ct=\rho-a$, rising past unity for all ρ, and then falling to a ρ-independent limit when $ct>>\rho+a$.

APPENDIX D - EHRENFEST'S THEOREM

Consider a monoenergetic beam of electrons incident from the +x direction and moving in the -x direction toward an excluded cylinder of radius a, centered on the z axis as in Fig. 2. The magnetic flux Φ is confined to radius $\rho < (a/2)$, so the electrons never exprience a local magnetic field. The Hamiltonian for this problem is given by

$$H = \frac{1}{2} mv^2 = - \frac{\hbar^2}{2m} (\nabla + \frac{ie}{\hbar c} A)^2 \tag{D.1}$$

with the boundary condition that $\psi = 0$ for $\rho = a$. The incoming wave is given by

$$\lim_{x \to +\infty} \psi \sim \exp\{-i\frac{\theta}{2} - ikx - i\omega t\} \tag{D.2}$$

For the case a=0, $\Phi = (1/2)\Phi_0$, Aharonov and Bohm [2] found that

$$\psi = \sqrt{\frac{i}{2}} \exp\{-i\frac{\theta}{2} - ikx - i\omega t\} \int_0^{\sqrt{k(\rho + x)}} \exp\{it^2\} dt \tag{D.3}$$

is an exact solution of the stationary Schroedinger equation in $-\pi \le \theta \le \pi$, and Ref. [4] confirmed that the limit $a \to 0$ is smooth. Here I shall use (D.3) to calculate the momentum transfer to the electron beam from the wall of the excluded cylinder in the limiting case a=0, and compare the result with the momentum transfer implied by the scattering cross section per unit length of cyclinder,

$$\frac{d\sigma}{d\theta} = \frac{1}{2\pi k \cos^2\left(\frac{\theta}{2}\right)} \tag{D.4}$$

which follows from (D.3).

The force on the electron beam is given by

$$F_x = \frac{d}{dt}\langle mv_x \rangle = \int \left[\frac{\partial \psi^*}{\partial t} mv_x \psi + \psi^* mv_x \frac{\partial \psi}{\partial t} \right] dxdy$$

$$= \frac{i}{\hbar} \int \left[(H\psi)^* mv_x \psi - \psi^* H(mv_x \psi) \right] dxdy + \frac{i}{\hbar} \int \psi^* [H, mv_x] \psi dxdy , \tag{D.6}$$

where

$$mv_x = \frac{\hbar}{i} \frac{\partial}{\partial x} + \frac{e}{c} A_x = \frac{\hbar}{i} \left(\frac{\partial}{\partial x} - \frac{i}{2\rho} \sin\theta \right) \tag{D.7}$$

The second integral in (D.6) leads to the expectation of the Lorentz force, $(-e/c)v \times B$, in the usual way, and vanishes in the AB case because B vanishes on the support of ψ. The first term may be integrated by parts to give

$$\frac{d}{dt}\langle mv_x \rangle = \lim_{\rho \to 0} \left\{ \frac{i\hbar}{2m} \int_{-\pi}^{\pi} \rho d\theta \left[\frac{\partial \psi^*}{\partial \rho} mv_x \psi - \psi^* \frac{\partial}{\partial \rho} (mv_x \psi) \right] \right\} , \tag{D.8}$$

where I have omitted the surface integrals at infinity because I am only interested in the force at the surface of the cylinder. In a proper treatment with wave packets, the integrals at infinity would vanish in any case. For small $k\rho$,

$$\psi \to \sqrt{\tfrac{i}{2}} \sqrt{k(\rho+x)} \, \exp\{-i\tfrac{\theta}{2} - ikt - i\omega t\} \tag{D.9}$$

and (D.8) then gives

$$\frac{d}{dt}\langle mv_x \rangle = \frac{\hbar^2 k}{8m} \int_{-\pi}^{\pi} (1+\cos\theta)\, d\theta = \frac{\pi\hbar^2 k}{4m} \ . \tag{D.10}$$

Equation (D.10) should be compared with the momentum transfer rate,

$$\frac{d}{dt}\langle mv_x \rangle = \rho_i \left(\frac{\hbar k}{m}\right) \int_{-\pi}^{\pi} d\theta \left(\frac{d\sigma}{d\theta}\right) \hbar k (1+\cos\theta) \tag{D.11}$$

where the incident beam density ρ_i is determined from Eq. (D.3) as

$$\rho_i = \frac{\pi}{8} \ , \tag{D.12}$$

and the cross section is given by (D.4). That calculation yields

$$\frac{d}{dt}\langle mv_x \rangle = \frac{\hbar^2 k}{8m} \int_{-\pi}^{\pi} d\theta = \frac{\pi\hbar^2 k}{4m} \ , \tag{D.13}$$

in agreement with (D.10). Thus, Ehrenfest's theorem is obeyed and the needed force is provided by reflection from the boundary of the cylinder in this case. A formal proof for the general case with a finite excluded region can be obtained in the same way. One can also include the momentum transfer through a sphere at infinity, or use wave packets.

REFERENCES

[1] W. Ehrenberg and R. W. Siday, Proc. Phys. Soc. London B62 (1949) 8.

[2] Y. Aharonov and D. Bohm, Phys. Rev. 115 (1959) 485.

[3] W. H. Furry and N. F. Ramsey, Phys. Rev. 118 (1960) 623.

[4] M. Peshkin, I. Talmi, and L. J. Tassie, Ann. Phys. 16 (1960) 426.

[5] Y. Aharonov and D. Bohm, Phys. Rev. 123 (1961) 1511.

[6] B. S. DeWitt, Phys. Rev. 125 (1962) 2189.

[7] T. T. Wu and C. N. Yang, Phys. Rev. D 12 (1975) 3845.

[8] C. N. Yang, Proc. of the International Symposium "Foundations of Quantum Mechanics", Tokyo, 1983, ed. S. Kamefuchi et al. (Physical Society of Japan, 1984) p. 5.

[9] P. Bocchieri and A. Loinger, Lett. Nuovo Cimento 39 (1984) 148, and references therein.

[10] D. H. Kobe and J. W. Liang, Phys. Lett. 118A (1986) 475.

[11] J. Q. Liang and X. X. Ding, Phys. Lett. A119 (1987) 325, and references therein.

[12] W. Henneberger, J. Math. Phys. 22 (1981) 116, and references therein.

[13] W. Henneberger, Phys. Rev. Lett. 52 (1984) 573.

[14] B. Liebowitz, Nuovo Cimento 38 (1965) 932.

[15] T. H. Boyer, Phys. Rev. D 8 (1973) 1667 and 1679.

[16] e.g., M. Bawin and A. Burnel, J. Phys. A16 (1983) 217; A18 (1985) 2123.

[17] M. Peshkin, Phys. Rev. A 23 (1981) 360.

[18] R. G. Chambers, Phys. Rev. Lett. 5 (1960) 3.

[19] G. Mollenstedt and W. Bayh, Phys. Blatter 18 (1962) 299; Naturwiss. 4 (1962) 81.

[20] A. Tonomura, the following paper, and references therein.

[21] P.A. Horvathy, Phys. Rev. D 33 (1986) 407.

[22] R. Sundrum and L. J. Tassie, J. Math. Phys. 27 (1986) 1566.

[23] F. Wilczek, Phys. Rev. Lett. 48 (1982) 1144.

[24] R. A. Webb, S. Washburn, A. D. Benoit, C. P. Umbach, and R. B. Labowitz, Proc. of the 2nd International Symposium "Foundations of Quantum Mechanics", Tokyo, 1986, ed. M. Namiki et al, (Physical Society of Japan, 1987), p. 193, and references therein.

[25] M. Buttiker, New Techniques and Ideas in Quantum Measurement Theory, (New York, 1986), ed. D. M. Greenberger (Annals of the New York Academy of Sciences, 1986), p. 194.

[26] D. M. Greenberger, D. K. Atwood, J. Arthur, C. G. Shull, and M. Schlenker, Phys. Rev. Lett. 47 (1981) 751.

[27] T. Matsuda, S. Hasegawa, M. Igarashi, T. Kobayashi, M. Naito, H. Kajiyama, J. Endo, N. Osakabe, A. Tonomura, and R. Aoki (Phys. Rev. Lett., to be published).

[28]   S. Olariu and I. I. Popescu, Rev. Mod. Phys. 57 (1985) 339.

[29]   L. J. Tassie, Phys. Lett. 5 (1963) 43.

[30]   M. Peshkin, Phys. Rep. 80 (1981) 376.

[31]   H. Miyazawa (private communication) has also emphasized this important distinction.

[32]   L. J. Tassie and M. Peshkin, Ann. Phys. 16 (1961) 177.

[33]   M. J. Bowick, S. B. Giddings, J. A. Harvey, G. T. Horowitz, and A. Srominger, Phys. Rev. Lett. 61 (1988) 2823, and references therein.

[34]   H. J. Lipkin and M. Peshkin, Fundamental Aspects of Quantum Theory, ed. V. Gorini and A. Frigerio (Plenum Press, 1986), p. 295.

[35]   Y. Aharonov, Proc. of the International Symposium "Foundations of Quantum Mechanics", Tokyo, 1983, ed. S. Kamefuchi et al. (Physical Society of Japan, 1984), p. 10.

The Aharonov-Bohm Effect

Part Two: Experiment

Akira Tonomura

Advanced Research Laboratory, Hitachi, Ltd.
Kokubunji, Tokyo, Japan

1. INTRODUCTION

Most physicists have assumed for some time that electro-magnetism is completely described by electromagnetic fields. No physical influence has been thought to act upon a charged particle when it passes through a field-free region.

In 1959, however, Y. Aharonov and D. Bohm [1] predicted a paradoxical phenomenon, which later came to be called the Aharonov-Bohm, or AB effect. This effect describes how electrons are physically influenced, in the form of a phase shift, by electro-magnetic fields which the electrons do not experience. Though not conceivable within the framework of classical electrodynamics, Aharonov and Bohm actually solved the Schrödinger equation to show that this is the case in quantum mechanics. Soon after their prediction, several impressive experiments were carried out to demonstrate the existence of the effect.

In the mid-1970s, the significance of the AB effect to fundamental physics increased greatly due to the fact that the theory of gauge fields formulated in the 1950s was revived as the most probable candidate for a unified theory of fundamental physical interactions. In this theory, gauge fields are regarded as fundamental physical entities, just as their name indicates. Direct evidence of their physical reality is provided by the AB effect [2]. According to this view, it is the gauge fields, or vector potentials in this particular case, that exert a physical influence on electrons in a field-free region to produce the AB effect.

In 1978, controversy flared up concerning the existence of the AB effect. Bocchieri and Loinger [3] asserted that the AB effect is purely a mathematical concoction. They also questioned previous experiments, claiming that electrons must have been affected by fringing magnetic fields from finite whiskers or solenoids. A large number of papers were subsequently written most against this assertion. The number of papers concerned with this controversy has since risen to over 300.

In the attempt to settle things, several new experiments were carried out (see Chapter 5). Controversy surrounding the AB effect has entered a new phase.

Reviewing the long history of the AB effect controversy is not easy. A variety of different assertions have been published

concerning numerous aspects of the AB effect over the past 30 years.
I will do my best, in this part, to reproduce the various opinions.
I will try to introduce them without prejudice, from the standpoint
of an experimentalist. A clear theoretical discussion in Part One is
given by M. Peshkin, who has continued to make considerable
contributions to the theoretical discourse of the AB effect.

Following the process of the controversy will, I hope, be both
instructive and perhaps stimulating, especially for those who like
not only to study completed theory but to trace embarkations toward
uncharted worlds. Students may see from the history of the
controversy that problems cannot always be solved as in school
textbooks, but that working scientists are often confronted with a
seemingly endless series of unsolved problems.

The theme of this review is also pertinent to the study of
general physics, since the AB effect is deeply rooted in the
foundations of quantum mechanics. To be more precise, fundamental
topics in quantum mechanics such as the reality of vector potentials,
the locality of physical effects, the single-valuedness of the wave
function in a multiply connected region, and problems of magnetic
monopoles, are closely related to the AB effect. Thanks to the
controversy over the AB effect, their significance has for the first
time been thoroughly discussed in relation to numerous experiments.

I personally have no doubt that interest in the AB effect is
not a mere fad. Rather, it can provide us with deep insights into
the foundations of physics, while pointing in the direction of
diverse future developments.

2. AHARONOV-BOHM EFFECT

In 1959, Aharonov and Bohm presented a paper entitled "Significance of Electromagnetic Potentials in Quantum Theory" [1]. Its content can be roughly summarized as follows.

In classical electrodynamics, potentials are merely a convenient mathematical tool for calculations concerning electromagnetic fields. The fundamental equations can always be set up using fields. In quantum mechanics, however, potentials cannot be eliminated from the Schrödinger equation and consequently seem to have physical significance. The authors went beyond conjecture by proposing actual electron interference experiments. These experiments would be aimed at clarifying how potentials would affect electrons passing through field-free regions. The phenomenon the two authors described came to be called the Aharonov-Bohm effect in their honor.[*]

2.1 Electric AB Effect

The AB effect has a twofold nature; it can be classified broadly into electric and magnetic effects. Let's first take a look at the electric AB effect, which is schematically illustrated in Fig. 2.1.

An electron wave packet is split in two. Each of the two halves passes through a separate long metal cylinder, and the two packets overlap to form an interference pattern. When the potentials of the cylinders are zero, an interference pattern is observed which is determined only by the difference in length of the two paths.

---

[*] The existence of the magnetic AB effect was actually first predicted by Ehrenberg and Siday [4] in 1949. They mathematically formulated electron optics in terms of a refractive index represented by scalar and vector potentials, and described the magnetic effect later to become known as the AB effect as one application result. They found it very curious that a phenomenon associated with a steady flux, not a change in flux, would come into play.

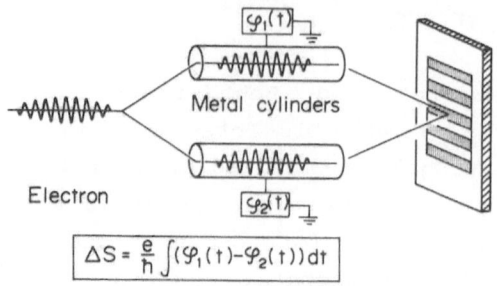

Fig. 2.1 Electric Aharonov-Bohm effect.

Time-dependent potentials, $\varphi_1(t)$ and $\varphi_2(t)$, are then applied to the two cylinders only when the electrons are well inside the cylinders. Since electric fields always vanish well within the cylinders, no forces are exerted upon the electrons. However, a quantum-mechanical calculation shows that a relative phase shift is produced between the two packets. The value of this shift, ΔS, is given by

$$\Delta S = \frac{e}{\hbar} \int (\varphi_1(t) - \varphi_2(t))dt, \quad \dots\dots\dots\dots\dots\dots\dots\dots (2.1)$$

where $\hbar$ is Plank's constant divided by 2π and e is the absolute value of the electron charge. Here, MKS units are employed for experimental convenience.

The derivation of this equation is as follows. The Schrödinger equation for this system is

$$i\hbar \frac{\partial \Psi}{\partial t} = (H_o - e\varphi)\Psi, \quad \dots\dots\dots\dots\dots\dots\dots\dots\dots\dots (2.2)$$

where H_o is the Hamiltonian in the absence of potentials.

In the case of zero potentials ($\varphi_1(t) = \varphi_2(t) = 0$), the overall wave function can be represented as the superposition of the wave functions for two unperturbed packets, 1 and 2, such that

$$\Psi_o = \Psi_1{}^o(\mathbf{r},t) + \Psi_2{}^o(\mathbf{r},t). \quad \dots\dots\dots\dots\dots\dots\dots\dots (2.3)$$

When potentials $\varphi_1(t)$ and $\varphi_2(t)$ are applied to the cylinders, the wave function becomes

$$\Psi = \Psi_1{}^0(r,t)e^{iS_1} + \Psi_2{}^0(r,t)e^{iS_2}, \quad \dots\dots\dots\dots\dots \quad (2.4)$$

where $S_i = \frac{e}{\hbar} \int \varphi_i(t)dt$ (i=1,2). Consequently the relative phase shift, ΔS, between the two packets is given by equation (2.1).

2.2 Magnetic AB Effect

Now for the magnetic AB effect. Integral $\int \varphi_i(t)\,dt$ can be interpreted as a component of the covariant product of two four-vectors, potentials $A_\mu = (A, -\varphi/c)$ and space-time differential dx^μ = (ds,cdt). Here, **A**, c, and **s** are respectively the vector potential, light velocity, and displacement vector. This generalization leads to the deduction that a magnetic AB effect should also exist.

The experimental configuration for this effect is shown in Fig. 2.2.

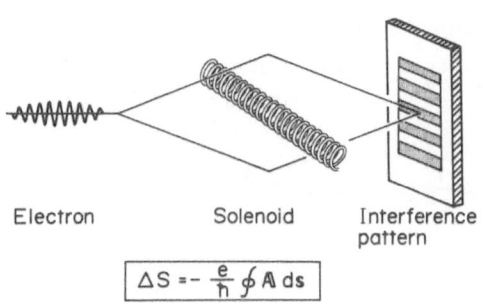

Electron   Solenoid   Interference pattern

$$\Delta S = -\frac{e}{\hbar} \oint A\, ds$$

Fig. 2.2 Magnetic Aharonov-Bohm effect.

When an electric current is applied to a closely wound solenoid, a magnetic field is produced only within the solenoid. An electron wave is split into two coherent waves. They pass on opposite sides of the solenoid and then recombine. Although there are no magnetic fields outside the solenoid, a relative phase shift between the two waves can be observed as an interference pattern. The phase shift, ΔS, in this case is given by

$$\Delta S = -\frac{e}{\hbar} \oint A\, ds, \quad \dots\dots\dots\dots\dots\dots\dots\dots\dots\dots \quad (2.5)$$

Here, the integral is carried out along a closed curve connecting the two paths.

Although magnetic field **B** is zero everywhere outside the solenoid, vector potential **A** cannot vanish there. This is because the loop integral of **A** around the solenoid is equal to magnetic flux $\Phi_0 = \int$ **B** d**S** inside it. Therefore, a nonzero phase shift can be observed. It should also be noticed that the phase shift remains unchanged under the gauge transformation **A** → **A** + grad χ.

Equation (2.5) can be derived in a manner quite similar to the electric case. The wave function is given by $\Psi(\mathbf{r},t) = \Psi_1^{\,O}(\mathbf{r},t) + \Psi_2^{\,O}(\mathbf{r},t)$ when there is no magnetic flux inside the solenoid. When the magnetic flux is nonzero, the wave function becomes $\Psi(\mathbf{r},t) = \Psi_1^{\,O}(\mathbf{r},t)e^{iS_1} + \Psi_2^{\,O}(\mathbf{r},t)e^{iS_2}$. Here $S_i = -\frac{e}{\hbar}\int \mathbf{A}\ d\mathbf{s}$, the line integral being carried out along each path. Therefore, the relative phase shift, ΔS, between the two beams is given by $S_1 - S_2 = -\frac{e}{\hbar}\oint \mathbf{A}\ d\mathbf{s}$, where the line integral is taken along a closed path determined by the two paths.

An exact solution for the problem of the scattering of electrons by a solenoid was also calculated by Aharonov and Bohm for the vanishing limit of the solenoid radius as follows (see Fig. 2.3). The Schrödinger equation for cylindrical coordinates is

$$\left[\frac{\partial^2}{\partial r^2} + \frac{\partial}{r\partial r} + \frac{1}{r^2}\left(\frac{\partial}{\partial\theta} + i\alpha\right)^2 + k^2\right]\Psi = 0, \quad \ldots\ldots\ldots\ldots \quad (2.6)$$

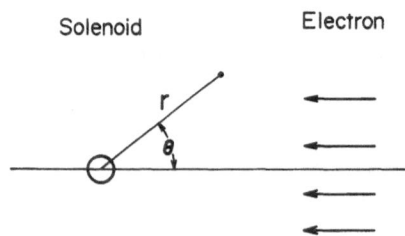

Fig. 2.3 Scattering of electrons by a solenoid.

Here, k and α are respectively a wave vector given by $\frac{2\pi}{\lambda}$, and the magnetic flux in the solenoid measured in h/e units, that is $\frac{e\phi}{h}$. The vector potential employed is such that $A_r = 0$ and $A_\theta = \phi/2\pi r$.

The solution of equation (2.6) can be determined under conditions where the current density of the incident electron beam is constant in the x direction. The incident wave function, then, is

$$\Psi_{inc} = e^{-i(kx + \alpha\theta)}. \quad\dots\dots\dots\dots\dots\dots\dots\dots\dots\dots\dots \quad (2.7)$$

This function is, in general, multi-valued when $\theta \to \theta + 2\pi$. This fact produces no problem since it holds only for the domain to the right of the origin. The total wave function, on the other hand, is single-valued everywhere. However, the "multi-valued" aspect of the incident wave function was later taken up as a point of controversy by opponents of the AB effect.

Aharonov and Bohm obtained a wave function in the asymptotic region, which takes the form

$$\Psi \longrightarrow e^{-i(kx + \alpha\theta)} + f(\theta)\frac{e^{ikr}}{\sqrt{r}}. \quad\dots\dots\dots\dots\dots\dots\dots \quad (2.8)$$

The cross section, σ, was given by

$$\sigma = |f(\theta)|^2 = \frac{\sin^2 \pi\alpha}{2\pi} \frac{1}{\cos^2 \frac{\theta}{2}} \quad\dots\dots\dots\dots\dots\dots \quad (2.9)$$

When α is an integer, the wave function can be obtained precisely as $e^{-i(kx + \alpha\theta)}$, and therefore $\sigma = 0$. The exact solution can also be obtained for $\alpha = n + \frac{1}{2}$, as given by

$$\Psi = \frac{i^{\frac{1}{2}}}{\sqrt{2}} e^{-i(\frac{\theta}{2} + kx)} \int_0^{\sqrt{kx(1+\cos\theta)}} \exp(iz^2)dz. \quad\dots\dots\dots\dots \quad (2.10)$$

The reader may already have noticed by considering the covariant product $A_\mu dx^\mu$ of four-vectors that the magnetic and electric effects are not two different phenomena. For example, the magnetic AB effect can also be observed as an electric effect in a coordinate system where the incident electron is at rest. It was shown by Lenz [5] that the invariant quantity to a Lorentz transformation can be given by

$$\oint A_\mu dx^\mu = \oint (\mathbf{A} \, d\mathbf{s} - \varphi \, dt), \quad\dots\dots\dots\dots\dots\dots\dots \quad (2.11)$$

which is an electromagnetic flux.

2.3 Significance of AB Effect

A strange conclusion necessarily follows from the considerations described up to now. In the case of the magnetic AB effect, the phase shift is determined only by the magnetic flux enclosed by the two paths, irrespective of whether the electrons experience magnetic fields or not. In order to understand how truly inconsistent this is with classical thinking, consider Young's double-slit experiments with electrons [6] using Fig. 2.4.

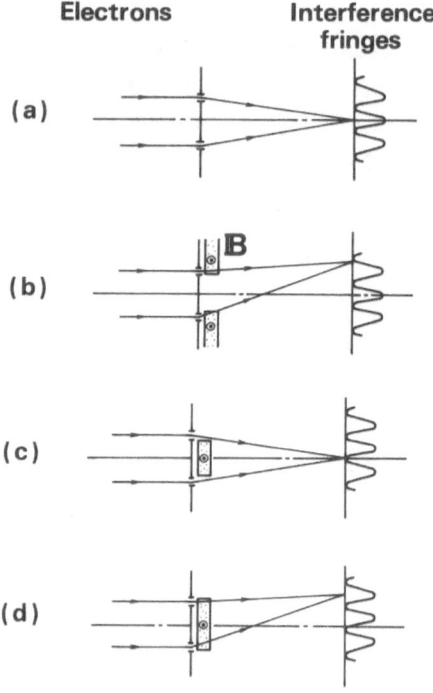

Fig. 2.4 Double-slit experiments with electrons.

A collimated electron beam is incident onto a screen with double slits and is diffracted. An interference pattern between the two diffracted beams is formed at the observation plane and can be recorded on film. Only the central part of the fringes is drawn in the figure. In the standard arrangement (a), the interference pattern shows its peak intensity at the central axis since the two beams are in phase at the intersection of the plane and the axis.

Up to here, there is no difference between this case and Young's experiment with light. Yet, what happens when we place a ferromagnetic thin film just behind the screen?

First, imagine two thin sheets of film placed on a plane located behind the screen. There is a gap between the sheets corresponding to the space between the slits in the screen. The film, however, covers the portions of the plane through which the beams pass (see Fig. 2.4(b)).

Since both beams are deflected at a small angle by the magnetic field, we might expect that the interference pattern is displaced as a whole. This means that the pattern should not necessarily have a peak intensity at the central axis. However, the fact is that the peak intensity does remain at the axis! This is because the relative phase shift between the two beams is completely determined by the enclosed magnetic flux, irrespective of the electron beam deflection.

The AB effect corresponds to case (c) in the figure. Although the electron beams do not touch the enclosed magnetic flux, a phase difference of $e\Phi/\hbar$ is produced between them, and the interference fringes are displaced. Furthermore, even when the electron beams actually cross the magnetic film at its edges and are deflected (see Fig. 2.4(d)), the interference pattern is left unchanged.

In classical mechanics, only the forces acting on electrons affect their behavior. Consequently, cases (a) and (c) are equivalent. Similarly, cases (b) and (d) are also equivalent. In quantum mechanics, however, the concept of forces gives way to that of electromagnetic potentials, and the potentials shift the electron phase. Here, cases (a) and (b) are equivalent, as are (c) and (d), though the paired cases are classically different.

We come here to a big question. Why does an electron notice the existence of the hidden magnetic field, when it never touches the magnetic field? Aharonov and Bohm discussed this problem as follows.

Electromagnetic potentials were only a mathematical aid in classical mechanics. In quantum mechanics, however, this is not the case. They believed the AB effect cannot be interpreted merely as a result of the local interaction of an electron with electromagnetic fields. If the local interaction principle coming from relativistic requirements is adhered to, there is no choice but to take the potentials to be more fundamental physical entities than the fields.

3. EARLY EXPERIMENTS

Predictions of the AB effect soon gave rise to the question of whether the existence of such a phase shift would have negated the electron interference fringes observed by Marton et al.[7]. This question was raised due to the fact that a weak, stray AC magnetic field was unintentionally included as part of the experimental conditions. However, Werner and Brill [8] settled this question by demonstrating via calculation that the relative phase shift due to the bending of interfering electron beams in a uniform magnetic field is perfectly compensated for by the phase shift due to the AB effect.

In 1960, the first electron interference experiment aimed at proving or disproving the AB effect was carried out by Chambers [9]. The predicted phase shift was detected using a tapered iron whisker instead of a solenoid. Subsequently, Fowler et al. [10] observed the phase shift using a slightly different arrangement, whereby the shape of the iron whisker could simultaneously be viewed as a de-focused electron microscopic image. Then, Möllenstedt and Bayh [11] observed the AB phase shift by changing the magnetic flux within a thin solenoid, and succeeded in encompassing all dynamic behavior within one photograph. Boersch et al. [12] also observed the phase shift using a Permalloy thin film evaporated onto a filament.

Superconducting interference was utilized instead of electron interference to demonstrate the AB effect by Jaklevic et al. [13]. Later, Matteucci and Pozzi [14] confirmed the AB phase shift using a biprism filament covered with a ferromagnetic layer.

Before moving on, it would be good to take a closer look at the reports by Chambers, and Möllenstedt and Bayh. Details of these experiments are also given in a review paper by Olariu and Popescu [15].

3.1 Chambers Experiment [9]

Chambers' experiment was the first to focus specifically on the AB effect. A transmission electron microscope was utilized to carry out an electron interference experiment on the AB effect.

A conceptual diagram of the experiment is given in Fig. 3.1(a). An electron beam from a point source is incident on a Möllenstedt type electron biprism [16], which is composed of a central fine filament and a ground-potential electrode on either side.

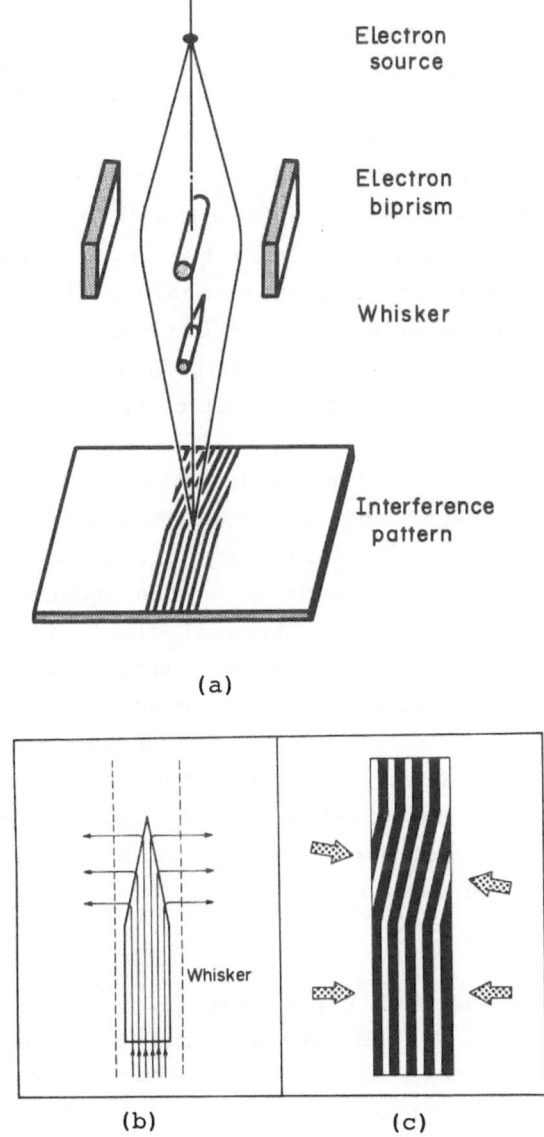

Electron
source

Electron
biprism

Whisker

Interference
pattern

(a)

Whisker

(b) (c)

Fig. 3.1 Chambers experiment using tapered iron whisker:
(a) Electron-optical system, (b) magnetic lines
of force around whisker, and (c)inter-
ference fringes.

When a positive potential of ∿10V is applied to the central filament, electrons passing on both sides of the filament are attracted towards the center. They then overlap to form an inter- ference pattern in the lower plane. The fringes are parallel to the filament. However, when a thin iron whisker ∿1μm in diameter was located in the shadow of the filament, the fringes were tilted where the whisker tapered (see Fig. 3.1(b) and (c)). Since magnetization inside the whisker is constant in the axial direction, the magnetic flux inside the whisker is proportional to the cross-sectional area.

The obtained interference pattern could clearly be explained as resulting from a phase shift proportional to the enclosed magnetic flux. The fringe tilting can also be explained classically as follows.

Since magnetic fields leak outside perpendicularly to the surface of the tapered whisker, electrons are deflected slightly in the axial direction. The deflection direction is, however, reversed on opposite sides of the whisker. Consequently, the two electron beams do not meet each other perpendicularly to the biprism filament in the observation plane. Rather, they are somewhat skewed (see Fig. 3.1(c)). As a result, the fringes tilt. This fringe tilting can thus be explained in terms of magnetic fields. However, the fringes are continuously connected from the tapered region to the uniform region of the whisker, which leads to the conclusion that the phase shift exists even in the uniform region. This phase shift cannot be explained classically.

3.2 Möllenstedt and Bayh Experiment [11], [17], [18]

An experiment by Möllenstedt and Bayh was both elegant and physically significant. According to Aharonov and Bohm [19], experi- ments up to this time had not been ideal in that the effect of vector potentials was mixed up with that of magnetic fields. In this experiment, however, special attention was paid to the detection of a pure potential effect.

A conceptual diagram of the experiment is shown in Fig. 3.2(a). A collimated electron beam is split by a biprism into two parts.

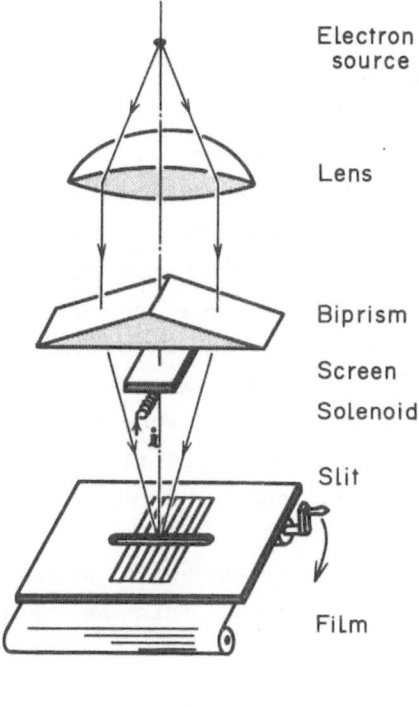

Electron
source

Lens

Biprism

Screen

Solenoid

Slit

Film

(a)

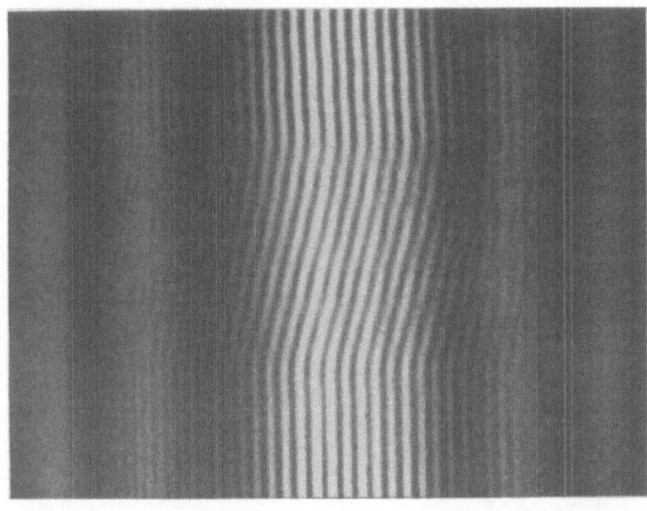

(b)

Fig. 3.2 Möllenstedt and Bayh experiment: (a)
          Electron-optical system, and (b) inter-
          ference pattern.

Then, they are again overlapped to form an interference pattern. An extremely thin solenoid [20] is located at the mid-point between the two beams. In the actual experiment, three electron biprism stages were provided so that the two beams might be kept far enough apart (∿ 120μm) as not to illuminate the solenoid. In addition, a ferromagnetic yoke was provided to remove the effect of fringing fields.

It was found that when the electric current applied to the solenoid was increased, the interference fringes moved laterally while the region of the whole interference pattern remained unchanged. In order to record this behavior, only a part of the interference pattern was recorded on film through a slit perpendicular to the fringe direction. When the film was made to move together with the increase in magnetic flux, dynamic behavior could be caught in the photograph shown in Fig. 3.2(b). While fringe tilting certainly resulted from the electric field induced by increasing magnetic flux, a fringe shift persisted even after the increase in magnetic flux stopped. This demonstrated the existence of the AB effect.

4. CONTROVERSY CONCERNING THE AHARONOV-BOHM EFFECT

Various kinds of arguments have occurred regarding the AB effect, ever since its prediction. Up to the mid-1970s, however, discussions mainly focused on theoretical interpretations of the AB effect. At that point, the significance of the AB effect increased greatly, for it came to be regarded as experimental evidence for the theory of gauge fields. As time progressed, however, even the effect's existence began to be questioned. Controversy flared afresh.

This chapter takes a historical approach to outlining these discussions.

4.1 Early Discussions

4.1.1 Quantum-Mechanical AB Effect

Widespread acceptance did not come at the time the AB effect was first predicted. Very few people had previously considered that electrons could be physically influenced by electric or magnetic fields when they were not directly touched by the fields. Electromagnetic potentials did appear in the Schrödinger equation, but they were thought to be a mathematical auxiliary and to produce no observable effect.

In 1960, an assertion was made in support of the AB effect by Furry and Ramsey [21]. This assertion was based on discussions of a complementarity principle whereby the AB effect is a fundamental phenomenon in quantum mechanics with no analogue in classical mechanics. The effect was seen to be essential to upholding the consistency of the theory. Wegener [22] pointed to the possibility of applying the AB effect to electron phase optics.

While prediction of the AB effect was made in the context of classical electromagnetic fields, actual fields experience vacuum fluctuations. In 1961, Mitler [23] investigated the effect of vacuum fluctuations on the measurability of the AB effect. He found that the AB effect was still observable even when fluctuations existed.

In 1964, Feynman et al. [24] took up the AB effect in their textbook "The Feynman Lectures on Physics". There, they explained the effect in lucid terms as evidence for the physical reality of vector potentials.

Peshkin, Talmi and Tassie [25] asserted that the AB effect is necessary for consistency with the uncertainty principle. They also predicted the validity of the AB effect for bound-state electrons orbiting around an isolated magnetic flux (see Fig. 4.1).

Magnetic flux : Φ

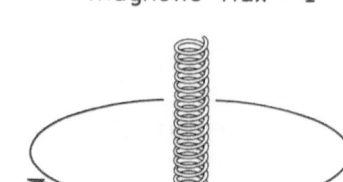

Electron

Fig. 4.1 Aharonov-Bohm effect for bound-state electrons.

According to their explanation, the kinetic angular momentum of the electron around the flux is influenced by the inaccessible flux. The angular momentum shifts from the allowed values, $m\hbar$ (m:integer), by $\dfrac{e\,\Phi}{2\,\pi}$. This occurs because the canonical angular momentum is quantized as $m\hbar$. They asserted from these results that the AB effect is not due to potential effects, but to modifications of the quantization conditions. Tassie and Peshkin [26] went on to show that the use of single-valued wave functions postulated for deriving the AB effect was justified in a cylindrically symmetric system according to the principles of symmetry theory.

Aharonov and Bohm [19] objected in their second paper to the interpretation of the AB effect presented by Peshkin, Talmi and Tassie [25]. They stated that, if the locality of the electro-magnetic interaction is properly reflected in the theory, it will be seen clearly that the magnetic field in the excluded region cannot directly interact with electrons without the intermediary of potentials.

In addition, they pointed out that a scattering cross section cannot be calculated from modifications of quantum conditions alone, without solving the Schrödinger equation using the specified vector potential outside the solenoid.

By this time, three experimental results had been reported that agreed with AB theory, and people seemed to believe in the existence of the AB effect. However, as Aharonov and Bohm stressed in their paper [19], none of the experiments could be regarded as an ideal confirmation. The significance of the potentials in quantum mechanics could not yet be considered to have been experimentally demonstrated.

4.1.2 Dispute Regarding Significance of Potentials

Subsequently, however, several people (Noerdlinger [27], DeWitt [28], Belifante [29], Feinberg [30], and Trammel [31]) presented arguments against Aharonov and Bohm's assertion that electromagnetic potentials have a physical significance in quantum mechanics. Noerdlinger [27] showed that the interaction of an electron with electromagnetic fields can be represented by fields alone if a theory non-local in time, whereby fields in the past can affect the present, is admitted.

In 1962, DeWitt [28] reformulated the theory of quantum electrodynamics utilizing electromagnetic fields alone without the use of the potentials. He concluded that there are no grounds for regarding potentials as a more fundamental physical quantity than fields. His argument is as follows.

A conventional wave function can be transformed to a gauge-invariant wave function, Ψ', such that

$$\Psi' = \exp\left\{-i\,\frac{e\Lambda}{\hbar}\right\}\Psi \quad\dotfill\quad (4.1)$$

by introducing gauge function Λ given below.

$$\Lambda = \int_{-\infty}^{O} A_\mu(z)\,\frac{\partial z^\mu}{\partial \xi}\,d\xi \quad\dotfill\quad (4.2)$$

where $z^\mu(x, \xi)$ $(\mu=1\sim4)$ are continuous functions of the space-time coordinates x^μ and a parameter, ξ, in the interval $-\infty < \xi \leq 0$. These functions vary from x^μ at $\xi=0$ to a spatial infinity at $\xi\to-\infty$ where the electromagnetic field vanishes. This wave function, Ψ', can be used in place of Ψ in the Schrödinger equation, provided that operator $\partial/\partial x^\mu$ is replaced by

$$\frac{\partial}{\partial x^\mu} + i\,\frac{e}{\hbar}\int_{-\infty}^{O} F_{\nu\sigma}(z)\frac{\partial z^\nu}{\partial \xi}\frac{\partial z^\sigma}{\partial x^\mu}\,d\xi \quad\dotfill\quad (4.3)$$

where

$$F_{\mu\nu} = \frac{\partial A\nu}{\partial x^{\mu}} - \frac{\partial A\mu}{\partial x^{\nu}}$$ .... (4.4)

DeWitt concluded that this formulation involves only field strengths, $F_{\mu\nu}$, though non-local line integrals do make an appearance. Consequently, he decided the AB assertion is false.

Aharonov and Bohm [32] responded to DeWitt's argument in 1962. They replied that non-local quantum electrodynamics without potentials is only a substitution by which the essential role of potentials is obscured. Potentials are eliminated in a trivial sense only, just as a linear equation can be asserted to be non-linear by substituting $y=z^2$ in the equation. An acceptable non-local formulation of electrodynamics should satisfy several requirements, they said, such as defining a complete set of observables for writing a wave function or expressing the mean value of a physical quantity.

Belifante [29], in 1962, further investigated DeWitt's formulation. DeWitt's wave function, Ψ', in equation (4.1) was seen not to be entirely gauge invariant, but rather to depend on the choice of integration path. This path dependence, however, could be removed by averaging DeWitt's gauge (equation (4.2)) over all directions of straight lines. The averaged gauge becomes a Coulomb gauge (div **A** =0), and the resulting theory can be written in terms of electric and magnetic fields, thereby achieving gauge-invariance. In addition, several complete sets of observables were defined which Aharonov and Bohm had presented as an integral part of the basic postulates of quantum theory. It was asserted by Belifante that this non-local formulation of quantum electrodynamics is superior to the conventional local theory in that it is based on a complete set of commuting observables.

Aharonov and Bohm [33] challenged Belifante's assertion in their fourth paper. They stated that, in order to achieve a truly local formulation without potentials, not just local basic operators, but also local interactions that do not involve non-local functions of the basic operators, are required.

Feinberg [30] did not agree with the assertion that the AB effect indicates a special role for potentials in quantum mechanics. In the electric AB effect (see Fig. 2.1), he postulated that the electron acquires an additional energy only within a cylinder where

all fields (and consequently forces on the electron) vanish. If we
take the electron out of the cylinder before the potential is
switched off, an energy shift can be observed. This energy arises
when an additional potential is applied to the cylinder; it exerts
influence on the frequency of the electron wave so as to shift the
phase.

Such information is recorded in the action, $\hbar S$. Consequently,
it cannot be observed classically. This is because classical
observables such as energy and momentum are represented by the
derivatives of hS, i.e. $-\hbar \frac{\partial S}{\partial t}$ and $\hbar$ grad S, respectively. In
quantum mechanics, however, the frequency and resulting phase
determine the absolute magnitude of $\hbar S$.

The situation is similar for the magnetic AB effect. No
observable effect can be classically produced on an electron passing
through a field-free region. Yet, this does not mean that there is
no physical interaction of the electron with the source of the
field, even in the classic sense. A moving electron produces a
magnetic field which acts on the solenoid to produce an interaction
energy represented by the e**A** term in the expression for total energy
$\frac{1}{2M}$ (p+e**A**)^2. As a result, the electron acquires an additional
action, $\frac{e\phi}{2}$, on one side of the solenoid, as well as $-\frac{e\phi}{2}$ on the
other side. This does not influence the electron motion nor add to
the total work the electron does on the solenoid.

In quantum mechanics, such a change in action appears as a
phase shift for each beam. This change, however, becomes observable
as a fringe shift only when the two beams enclose the solenoid and
overlap.

4.1.3 Classical Interpretation of AB Effect

Several trials were conducted as an attempt to interpret the AB
effect in the context of a classical interaction between the
incident electron and solenoid. Trammel [31] tried to assign the
magnetic AB effect to the classical interaction of an incident
electron with the solenoid, taking into consideration the force
exerted on the solenoid current by the electron.

Liebowitz [34] presented a rather drastic interpretation of the
magnetic AB effect in 1965. He stated that the AB effect could be
fully explained in terms of forces which so far have been overlooked.

Although this point of view is disputed by Peshkin (see Part
One, Chapter 5), details are given below for those concerned with

what this overlooked force might be (see Fig. 4.2)

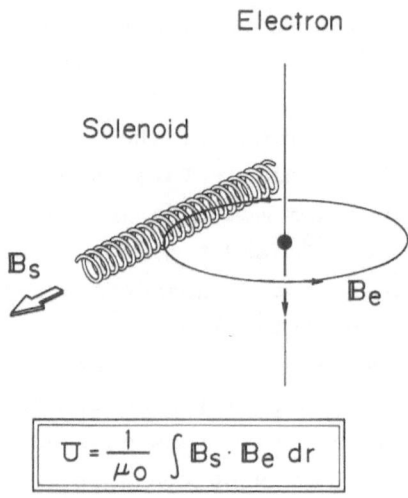

Fig. 4.2   Classical interpretation of Aharonov-Bohm
           effect by Liebowitz.

If the electron's magnetic field, $\mathbf{B_e}$, can penetrate into the
solenoid and overlap the solenoid's field, $\mathbf{B_s}$, the total magnetic
energy is given by

$$\int (\mathbf{B_e} + \mathbf{B_s})^2 \, dr/2\mu_0$$

having the interaction energy

$$U = \frac{1}{\mu_0} \int \mathbf{B_e} \mathbf{B_s} \, dr. \ \dots\dots\dots\dots\dots\dots\dots\dots\dots\dots\dots\dots \ (4.5)$$

Here, the integral is performed for the volume inside the solenoid.
     Both $\mathbf{B_e}$ and consequently U depend on the electron's position,
$\mathbf{r_e}$; and thus a force,

$$\mathbf{F} = - \text{grad } U(\mathbf{r_e}). \ \dots\dots\dots\dots\dots\dots\dots\dots\dots\dots\dots\dots \ (4.6)$$

is exterted on the electron. According to Liebowitz, this force
should be added to the Lorentz force - not only in the case of the
AB effect, but also in general - as follows:

$$F = -e(E + v \times B) - \text{grad } U \ (r_e) \dots\dots\dots\dots\dots \ (4.7)$$

Liebowitz stated that this new force term will vanish virtually always, even though in the present case it explains the AB effect. Electron motion is modulated a little in the longitudinal direction as it passes on one side of the solenoid, due to the force -grad $U(r_e)$. The force on the other side of the solenoid is the exact opposite. This is because the value of interaction energy, U, is reversed in sign between the two cases. The calculated difference in the two displacements corresponds exactly to the AB phase shift when de Broglie's relation is employed.

Against Liebowitz's argument, Hraskó [35] asserted that the Lorentz force already contains the contribution of a -grad U term (equation (4.7)) in the Lagrangian, and that a repeated addition to the Lorentz force cannot be justified. Liebowitz [36] in turn disputed this assertion. He put forward that the Lorentz force is, by definition, determined by fields at the electron's position, though the force given by -grad U is determined by fields inaccessible to the electron. The interaction energy, U, he said, cannot be used in a Lagrangian context in spite of its formal resemblance to a familiar term in a standard Lagrangian.

Against the classical interpretations by Liebowitz [34] and Trammel [31], Kasper [37], [38] argued that the basic assumption of an overlap of the electron's magnetic field with the solenoid's is invalid. This is because the magnetic field of the incident electron is shielded by the skin effect in the solenoid surface. Liebowitz [39] disputed Kasper's assertion, taking the arrangement of an actual solenoid into consideration when doing so.

Erlichson [40], reviewing the literature on the AB effect up to 1969, regarded the localizability principle in physical effects as the key problem regarding the AB effect. He proposed testing possible local interaction between the excluded field in the solenoid and the electron's field.

In his theoretical experiment, a superconducting barrier is placed between the electron and the solenoid in order to prevent the penetration of the electron's magnetic field into the solenoid. He stated that if the AB effect disappears due to superconducting shielding, the AB effect may be explained within the framework of a localizability principle for physical effects, even without ascribing a causal role to potentials. Furthermore, he expected from experimental evidence for magnetic flux quantization (Deaver and Fairbank [41]) that the AB effect would vanish under these circumstances.

However, Erlichson's interpretation was later seen not to be the case once the problem was experimentally tested by Lischke [42] in an electron interference experiment using a superconducting cylinder trapping a magnetic flux. A similar experiment was also performed by Wahl [43]. Erlichson's proposal was later confirmed in a more ideal experiment (see Chapter 5). In this experiment, a toroidal magnetic field was completely shielded by a superconducting layer.

In 1973, Boyer [44] classically investigated interaction of an electron with a solenoid, and verified conservation laws connected with energy and momentum that did not require introduction of any new force. He stated that Liebowitz's conclusion [34], whereby an incident electron receives a force from the solenoid, is erroneous. Since the solenoid current is kept constant, he said, some external forces must keep them moving without any disturbance. He reported that these forces do work, appearing as energy, ΔE, in the electromagnetic fields. Here, ΔE can be given by

$$\Delta E = -e\mathbf{v}\mathbf{A}, \dots\dots\dots\dots\dots\dots\dots\dots\dots\dots\dots\dots \quad (4.8)$$

where $\mathbf{A}$ is the vector potential in the Coulomb gauge of the solenoid at the electron's position. From this point of view, however, the solenoid does not exert force on the passing electron. This is because it is not the passing electron but the external forces that counteract it.

In a later paper by Boyer [45], however, a possible lag effect was predicted. According to this interpretation, if the energy change, ΔE, in the electromagnetic field can be assumed to be compensated for by a change in the kinetic energy of the passing electron, then the velocity change, Δv, is

$$\Delta v = \frac{e}{p} \, \mathbf{v}\mathbf{A}. \dots\dots\dots\dots\dots\dots\dots\dots\dots\dots\dots\dots \quad (4.9)$$

Here, p is the electron momentum.

According to this result, the electron slows when approaching the solenoid and then speeds up when leaving. On the other side this is reversed. When the relative lag between two beams passing on opposite sides of the solenoid is calculated and associated with a relative phase shift, a phase shift of $e\Phi/\hbar$ can be obtained, which is exactly the same as for the AB effect.

In spite of difficulties in arriving at a realistic account of the forces, Boyer explained these forces using a simplified calculation. His explanation goes like this: the electron passing by the solenoid produces a small acceleration on the electrons inside the solenoid wire, which in turn produce electric and magnetic fields back at the electron's position. He asserted that there seemed to be no conclusive experimental evidence to rule out the lag effect and consequently to support the significance of the potentials. Further, he proposed three kinds of experiments to make the necessary distinctions. One would involve observing the interference pattern by creating a phase shift larger than the coherence length of the electron beam. The other two would be used to detect small velocity changes by measuring the difference in both arrival time and electrostatic deflection of the electrons on opposite sides of the solenoid.

4.2 Increased Significance of AB Effect

4.2.1 Non-integrable Phase Factor

In the mid-1970s, interest was renewed in the physical significance of the AB effect. The theory of gauge fields, proposed in 1954 by Yang and Mills [46] and in 1956 by Utiyama [47], was taking on greater significance. Subsequent to the successful unification of electromagnetism and the weak force by Weinberg [48] and Salam [49], gauge field theory came to be regarded as the most probable candidate for a unified theory of all interactions.

In 1975, Wu and Yang [2] formulated a complete description of electromagnetism which focused on a new physical variable called "a non-integrable (path-dependent) phase factor". They generalized this formulation to a non-Abelian gauge field using the mathematical theory of fiber bundles. In their formulation, the non-integrable phase factor corresponds to a parallel transport in the fiber bundle formulation (see Bernstein and Philips [50], and Tourrenc [51]). In this context, the AB effect corresponds to a self-evident geometrical theorem; it is considered an experimental manifestation of the non-integrable phase factor.

The various assertions we have reviewed up to this point are concerned with the question of what is the most fundamental physical quantity in electromagnetism. Some have said it is the field strength, and others the potential. In 1975, Wu and Yang [2] pointed to a physical quantity which has a one-to-one correspondence with what we can experimentally observe in electromagnetism, but is neither the field strength, $F_{\mu\nu}$, nor the vector potential, A_{μ}. Rather they considered

$$\exp\left\{-\frac{ie}{\hbar}\oint A_{\mu}dx^{\mu}\right\} = \exp\left\{-\frac{ie}{\hbar}\left(\oint \mathbf{A}ds - \oint\varphi dt\right)\right\}. \quad \ldots\ldots\ldots(4.10)$$

Equations (2.1) and (2.5) point to the fact that this phase factor is exactly the same as for the AB effect experiment. The integral is carried out along a closed loop determined by two paths. When the integral is performed not along a loop, but along a path connecting two points, P and Q, then the quantity is not only a function of points, P and Q, but is path-dependent. Therefore, it is described as a non-integrable phase factor, like this:

$$\exp\left\{-\frac{ie}{\hbar}\int_{P}^{Q}A_{\mu}dx^{\mu}\right\}. \quad \ldots\ldots\ldots\ldots\ldots\ldots\ldots\ldots\ldots\ldots\ldots (4.11)$$

This phase factor is pictured in a practical AB-effect arrangement in Fig. 4.3.

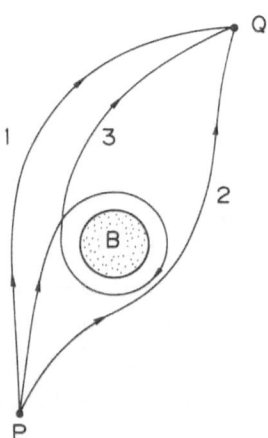

Fig. 4.3   Phase factors have different values
along paths 1,2 and 3.

The non-integrable phase factor is a complex number, and its value is different for each topologically different path. Paths 1 and 2 respectively pass on the left and right sides of the solenoid, and path 3 passes once around the solenoid in a clockwise direction.

This example makes it clear that the phase factor reflects the topological nature of its path. This non-integrable phase factor, however, is not observable, since there is no way to measure the phase shift along a path from P to Q. The only way to observe it is to look at the interference pattern created by the phase shift relative to another reference path from P to Q. What can actually be observed is then described by the phase factor whereby the integral is performed along the closed loop.

This situation is fully expressed by Wu and Yang [2] like this:

- The local field strength, $F_{\mu\nu}$, underdescribes electro-magnetism. This is proven by the AB effect.
- The phase $\frac{e}{\hbar} \oint A_\mu dx^\mu$ overdescribes electromagnetism, since no experiment can distinguish between two cases where the values are only different by multiples of 2π.
- Electromagnetism is completely described by the phase factor. This factor contains necessary and sufficient observable information about electromagnetism.

Amazingly enough, all electromagnetic phenomena, with all their various aspects, can then be expressed by the phrase, "Electro-magnetism is the gauge-invariant manifestation of a non-integrable phase factor."

4.2.2 Generalization to Non-Abelian Gauge Fields

Formulation of electromagnetism using a non-integrable phase factor was further generalized to non-Abelian gauge fields. The significance of the phase factor increases there, since the field strengths underdescribe the gauge field even in a simply-connected region.

Wu and Yang [2] proposed an extension of the AB effect that could be used as a definitive test for the existence of the isotopic spin gauge field, which describes both proton and neutron as different isospin states of a nucleon. They made this proposal because no experimental proof up to that time had required the existence of the gauge field. In the experiment they proposed, the solenoid used to create the gauge field would be a cylinder rotating around its axis, which would be made of heavy elements with an excess of neutrons (see Fig. 4.4).

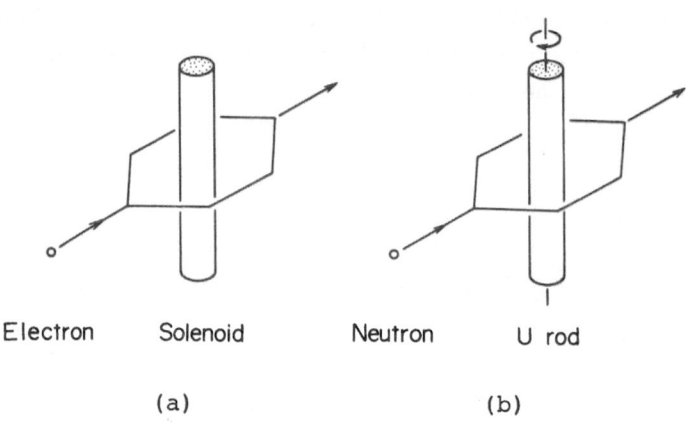

Electron   Solenoid   Neutron   U rod

(a)   (b)

Fig. 4.4   Schematic of generalized AB effect:
(a) Magnetic AB effect, and (b) generalized
AB effect.

The rotation of neutrons instead of electrons would form a nonzero field strength inside the cylinder. The fringe shift observed using a proton beam would be in the opposite direction from that using a neutron beam.

Such an experiment was later attempted by Zeilinger, Horne and Shull [52]. However, a positive result could not be obtained. This was attributed to the fact that the isospin AB-like interaction was estimated to be weaker than interaction due to the electron AB effect by a factor of 10^{-15}. The interpretation was that the range of the isotopic spin gauge field would be something like the pion Compton wavelength.

Further theoretical investigations concerning this extension of the AB effect were carried out by Zeilinger [53], Botelho and Mello

[54], Horváthy [55], and Sundrum and Tassie [56]. Although the predictions of Wu and Yang were qualitative, Horváthy [55] gave a more quantitative description. Sundrun and Tassie [56] studied the AB effect in the context of general gauge theories, using the Feynman path integral method [57] . They pointed out that the SU(2) AB experiment proposed by Wu and Yang would only test the presence of an Abelian subgroup of the gauge group. A true non-Abelian AB effect, they said, is required to indicate a non-Abelian gauge group.

The AB effect was also extended beyond the isotopic spin gauge field to the gravitational field. Studies were reported by many authors, including Wisnivesky and Aharonov [58], Dowker [59], Papini [60], Greenberger [61], Krauss [62], Overhauser and Collela [63], Anandan [64,65], Ford and Vilenkin[66], Stachel [67], Burges[68], Ferrari and Griego [69], Parthasarathy, Rajasekaran and Vasdevan [70], and Bezerra [71]. Aharonov and Vardi [72] predicted that the AB effect would work such that the enclosed magnetic flux could modify the average value of a particle's spin. However, Olariu and Popescu [15] asserted that there could be no effect on the spin from an enclosed flux. An analogue of the AB effect due to a long range interaction between spins was predicted for both electrons and photons by Naik [73] in 1986.

Although perhaps not entirely relevant to a generalized AB effect involving non-Abelian gauge fields, it may be good at this point to touch upon two particular investigations. One was an experimental test carried out by Greenberger et al. [74] to determine whether there is an AB effect for neutron beams. Since a neutron has no electric charge, existence of the AB effect would provide evidence for a breakdown of the standard minimal coupling. However, their neutron interference experiment using a toroidal ferromagnet gave a null result.

The other investigation concerned the AB effect for a neutral particle, as predicted by Aharonov and Casher [75] in 1984. The experimental arrangement was just the reverse of that for the magnetic AB effect (see Fig. 4.5).

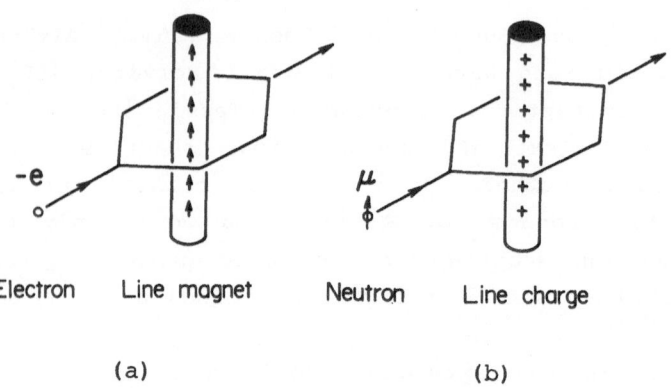

Electron   Line magnet   Neutron   Line charge

(a)   (b)

Fig. 4.5   Schematic of Aharonov-Casher effect:
(a) AB effect, and (b) AC effect.

In this case, a solenoid, i.e. a line of magnetic dipoles, was replaced by a line of electric charges. A neutron with a dipole moment acted as an electron.

Although the neutron beam was subjected to no force, a phase shift, ΔS, was produced between two beams enclosing the charged line:

$$\Delta S = \frac{\mu \lambda}{\hbar} \quad \dots\dots\dots\dots\dots\dots\dots\dots\dots\dots\dots \quad (4.12)$$

Here, u is the projection of the magnetic moment along the line, and λ is the charge density. This effect can be intuitively understood if we switch to the rest coordinate system for the neutron. A magnetic field arises from movement of the charged line, which exerts an influence on the magnetic moment to produce the phase shift (see Klein [76]).

4.3   Controversy Since the Mid-1970s

4.3.1   Nonexistence of AB Effect

In 1978 Bocchieri and Loinger [3] claimed the AB effect did not exist; controversy had thus spread even to the existence of the effect. They asserted that the AB effect is actually gauge dependent and a purely mathematical concoction. All consequences of quantum mechanics are dependent on field strengths and not on potentials, they said. Although they put forward many different reasons for this assertion, for reasons of space only the most salient are summarized here.

(1) Non-Stokesian vector potentials

According to their analysis, a gauge function can be chosen so that the vector potential **A** completely vanishes outside an infinite solenoid; consequently there is no AB effect [3,77]. The vector potential **A** in the Coulomb gauge around a solenoid with radius a, which contains magnetic flux Φ, can be given by (see Fig. 4.6(a))

$$A_r = A_z = 0, \qquad A_\theta = \frac{\Phi}{2\pi r} \qquad (r \geqq a)$$
$$A_r = A_z = 0, \qquad A_\theta = \frac{\Phi r}{2\pi a^2} \qquad (r < a).$$

$$\left. \right\} \quad \dots\dots\dots \quad (4.13)$$

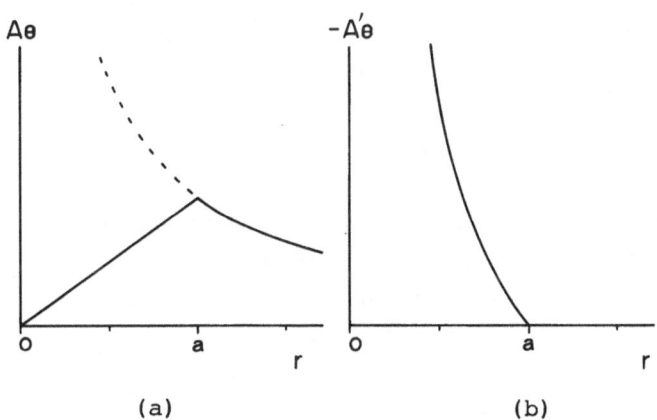

(a)   (b)

Fig. 4.6 Vector potentials around infinite solenoid:
(a) Coulomb gauge, and (b) Bocchieri and
Loinger's gauge.

Here, the vector potential vanishes outside the solenoid when gauge transformation

$$\Lambda = -\frac{\Phi\theta}{2\pi} \quad \dots\dots\dots\dots\dots\dots\dots\dots\dots\dots\dots\dots \quad (4.14)$$

is introduced. New vector potential **A'** (see Fig. 4.6(b)) then becomes

$$\mathbf{A'} = 0 \qquad\qquad (r \geq a)$$
$$\mathbf{A'}_r = \mathbf{A'}_z = 0, \qquad \mathbf{A'}_\theta = \frac{\Phi r}{2\pi a^2} - \frac{\Phi}{2\pi r}. \quad (r < a) \qquad \Bigg\} \quad \cdots \cdot (4.15)$$

Therefore, an electron passing outside of the solenoid is unaffected.

Later, Home and Sengupta [78] presented a different vector potential, **A"**, for an infinite solenoid, which also vanishes outside the solenoid. It can be given by

$$\mathbf{A"}_r = \mathbf{A"}_z = 0, \qquad \mathbf{A"}_\theta = \frac{\Phi r}{2\pi a^2} \quad \cdots\cdots\cdots\cdots\cdots (4.16)$$

inside the solenoid. Home and Sengupta preferred this new gauge which is well-defined and continuous at r=0, because rot **A'** is undefined at r=0 in the case of Bocchieri and Loinger's gauge. Although this vector potential is not continuous at the solenoid boundary and does not satisfy Stokes' theorem, they stated that the validity of Stokes' theorem was not an essential condition since the magnetic field itself was discontinuous across the boundary.

(2) Hydrodynamical formulation using field strengths alone

According to Bocchieri and Loinger [3,79], the Schrödinger equation can be replaced by a set of nonlinear differential equations called "hydrodynamical equations", which contain only field strengths **E** and **B**. Therefore there is no room for the AB effect. When wave function ψ is written $\Psi = \sqrt{\rho}\, e^{iS}$, the following equations can be derived for the electron density, $\rho(\mathbf{r},t)$, and velocity, $\mathbf{v}(\mathbf{r},t)$.

$$\text{div}(\rho\, \mathbf{v}) + \frac{\partial \rho}{\partial t} = 0$$
$$M\frac{d\mathbf{v}}{dt} = -e(\mathbf{E} + \mathbf{v} \times \mathbf{B}) - \frac{\hbar^2}{2M} \text{grad}\left(\frac{\Delta^2 \sqrt{\rho}}{\sqrt{\rho}}\right), \qquad \Bigg\} \cdots\cdots (4.17)$$

where $M\mathbf{v} = \hbar\, \text{grad}\, S + e\mathbf{A}$.

(3) AB effect for bound-state electrons

Bocchieri and Loinger asserted the nonexistence of the AB

effect for bound-state electrons as follows. The kinetic angular momentum and energy of a rotating electron around a solenoid (see Fig. 4.1) do not depend on the enclosed magnetic flux. This is because the flux dependence, i.e. the AB effect for bound-state electrons, can automatically be derived by a gauge transformation, thus indicating that the AB effect is a purely mathematical concoction [3].

The Schrödinger equation for an electron orbiting around an axis which contains zero magnetic flux is

$$\frac{1}{2Ma^2} \left(-i\hbar \frac{\partial}{\partial \theta} \right)^2 \Psi = E \Psi , \quad \dots\dots\dots\dots\dots \quad (4.18)$$

Here, M and a are the electron mass and the radius of the orbit, respectively. The single-valuedness condition, i.e. $\Psi(\theta + 2\pi) = \Psi(\theta)$, makes the eigenvalue of canonical angular momentum $-i\hbar \frac{\partial}{\partial \theta}$, quantized as $m\hbar$ (m: integer). The energy eigenvalue is

$$E = \frac{\hbar^2 m^2}{2Ma^2}. \quad \dots\dots\dots\dots\dots\dots\dots\dots \quad (4.19)$$

Bocchieri and Loinger introduced a new representation for the same physical state that has just been described. It makes use of the gauge function given by $-\Lambda(\theta)$ in equation (4.14). The resultant vector potential and Schrödinger equation are

$$A_r = A_z = 0, \qquad A_\theta = -\frac{1}{r} \frac{\partial \Lambda}{\partial \theta} = \frac{\Phi}{2\pi r} \quad \dots\dots\dots \quad (4.20)$$

and

$$\frac{1}{2Ma^2} \left(-i\hbar \frac{\partial}{\partial \theta} + \frac{e\Phi}{2\pi} \right)^2 \Psi = E\Psi . \quad \dots\dots\dots \quad (4.21)$$

If $\Lambda(\theta)$ is single-valued, i.e. $\Lambda(\theta) = \Lambda(\theta + 2\pi)$, the two Schrödinger equations should have the same eigenvalues given by equation (4.19). However, since $\Lambda(0)$ is not generally equal to $\Lambda(2\pi)$ as seen from equation (4.14), the energy eigenvalue of equation (4.21) is

$$E = \frac{\hbar^2}{2Ma^2} \left(m + \frac{e\Phi}{h} \right)^2 . \quad \dots\dots\dots\dots\dots \quad (4.22)$$

If we think of Φ in this equation as the enclosed magnetic flux, then it can be concluded that the AB effect is produced by a purely mathematical procedure, i.e. a gauge transformation.

(4) AB Scattering

In 1981, Henneberger [80] asserted that there was no AB scattering. Electron scattering by an infinite solenoid had first been calculated by Aharonov and Bohm [1], and concluded to exist unless $\alpha = \frac{e\phi}{h}$ is an integer. However, the derived wave function is not single-valued (see equations (2.8) and (2.10)) unless $\frac{e\phi}{h}$ is an integer. Moreover, the total cross section diverges. Henneberger attributed this difficulty to incorrect criteria (continuity and single-valuedness) imposed by Aharonov and Bohm for wave functions. He stated that wave functions satisfying the Pauli criteria - square integrability, and closure under the velocity operators - lead to well-defined, sensible physics with no AB scattering.

The "nonexistence" of AB scattering was also asserted by Liang [81]. Non-conservation of angular momentum in AB scattering was claimed by Kobe and Liang [82].

4.3.2 Dispute Concerning Nonexistence of AB Effect

A number of papers were presented against the arguments for nonexistence of the AB effect. There were also a few speaking for these anti-AB arguments. A dynamic series of ongoing discussions took the form of repeated criticisms and responses, making it difficult to review every detail. Therefore, I attempt merely to summarize them as they correspond to each item discussed just above.

(1) Non-Stokesian vector potentials

The conditions that would have to be satisfied by vector potentials had already been investigated in connection with the refractive index formulation of electron optics by Ehrenberg and Siday [4]. They defined the refractive index, μ, from Fermat's principle like this:

$$\delta \int_P^Q \mu \, ds = 0. \quad \ldots\ldots\ldots\ldots\ldots\ldots\ldots\ldots\ldots\ldots \quad (4.23)$$

They also imposed conditions on index μ, which can be expressed, for example in a purely magnetic case, as

$$\mu = 1 + \frac{\widehat{s}A}{BR} \quad \ldots\ldots\ldots\ldots\ldots\ldots\ldots\ldots\ldots\ldots \quad (4.24)$$

Here **B**, R, and **ŝ** are respectively the magnetic field, the radius of a circular electron trajectory in a uniform magnetic field **B**, and the unit vector along the electron path. They stated that, in order not to violate validity conditions for Fermat's principle, the refractive index μ should: (1) be fixed everywhere in space once fixed in the neighborhood of one point; (2) have no singularities that make integral (4.23) convergent; and (3) have only such discontinuities as those which appear as limiting cases of μ. Therefore, the vector potential must satisfy Stokes' theorem, which is the only valid restriction. Under these restrictions, they said, vector potentials cannot, in general, vanish with the magnetic field.

The vector potential proposed by Bocchieri and Loinger (equation (4.15)) does not satisfy Stokes' theorem and is consequently called non-Stokesian. The inadmissibility of non-Stokesian vector potentials were asserted by Klein [83], Zeilinger [84], Bohm and Hiley [85], Mignago and Novaes [86], as well as Bawin and Barnel [87]. Their assertions are as follows.

The two vector potentials **A** and **A'** (equations (4.13) and (4.15)) seem, at first sight, equivalent to each other, since they are related by a gauge transformation

$$\mathbf{A'} = \mathbf{A} - \text{grad} \left(\frac{\Phi \theta}{2 \pi} \right). \quad \ldots\ldots\ldots\ldots\ldots\ldots\ldots\ldots\ldots\ldots \quad (4.25)$$

However, the non-Stokesian vector potential **A'** does not really describe the physical situation of the infinite solenoid shown in Fig. 4.7(a).

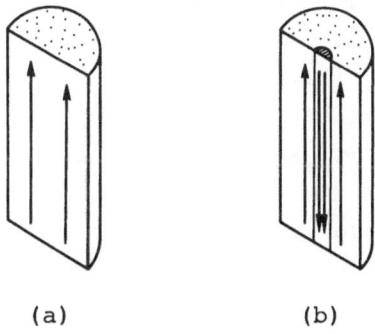

(a)   (b)

Fig. 4.7 Magnetic flux distributions represented by two kinds of gauges: (a) Coulomb gauge, and (b) Bocchieri and Loinger's gauge.

Rather, it describes the situation shown in Fig. 4.7(b). That is, an infinitely thin solenoid with magnetic flux - Φ is placed along the central axis, in addition to the original solenoid with magnetic flux Φ. This can easily be confirmed by calculating magnetic field $\mathbf{B}'$ from $\mathbf{A}'$ in equation (4.15). The resultant magnetic field is

$$\mathbf{B}' = \left\{ B - \Phi \delta(\mathbf{r}=0) \right\} \hat{z} \quad \dots\dots\dots\dots\dots\dots\dots\dots\dots \quad (4.26)$$

where $\hat{z}$ is a unit vector in the solenoid direction. In this way, the total magnetic flux inside the solenoid vanishes for a non-Stokesian vector potential. This is why the AB effect cannot be derived.

The above assertions were again disputed by Bocchieri and Loinger [88]. They remarked that an impenetrable solenoid can be described with the vector potential, even when the potential is undefined inside the solenoid. There is no need to define the vector potential in a region inaccessible to the electron, they said. However, if magnetic field $\mathbf{B}'$ at $\mathbf{r}=0$ is assumed to be

$$\mathbf{B}'(\mathbf{r}=0) = \lim_{r \to 0} \operatorname{rot} \mathbf{A}' = B_0 \, \hat{z}, \quad \dots\dots\dots\dots\dots\dots \quad (4.27)$$

the non-Stokesian potential can also describe the real situation for an infinite solenoid.

The vector potential $\mathbf{A}''$ (equation (4.16)) proposed by Home and Sengupta [78] was disputed by Henneberger [89] in the following way. This vector potential, he stated, is also non-Stokesian, and thus not allowable. The return flux is embedded in the wall of the solenoid, so that total flux is zero.

Burnel and Reekmans [90] found for vector potential $\mathbf{A}''$ an impermissible singularity of magnetic field $\mathbf{B}''$ at the wall. However, Home [91] stated again that since the solenoid has discontinuity of $\mathbf{B}''$ across the wall, the values of $\mathbf{B}''$ and $\mathbf{A}''$ at the wall are no longer definable concepts. Even the relation $\mathbf{B}''=\operatorname{rot} \mathbf{A}''$, he said, is not meaningful there.

(2) Hydrodynamical formulation using field strengths alone

A hydrodynamical description of quantum mechanics was first suggested by Madelung [92] and later formulated as a self-consistent scheme by Bohm [93] and Takabayasi [94]. The

AB effect was initially discussed in this formalism by Strocchi and Wightman [95]. Contrary to the position of Bocchieri and Loinger, they asserted that the wave function always has a tail penetrating into the region of a non-vanishing magnetic field. This tail then produces the AB effect by local interaction. They remarked that, although the AB effect supports the conclusion that the relative phase shift between the two electron waves can be produced without any time delay even in an infinitely distant region, there is no problem since Schrödinger-theory effects can be propagated instantly.

Bocchieri and Loinger denied the entire existence of the AB effect on grounds that the hydrodynamical equations (equation (4.17)) were described utilizing field strengths alone [3,77,79]. However, this assertion was disputed by Bohm and Hiley [85], and Takabayasi [96,97]. The AB effect, they felt, can be interpreted using field strengths in the form not of local interaction but of non-local interaction.

In their view of hydrodynamical formalism, the following non-local equation should be added to equation (4.17):

$$M \oint v \ ds = mh + e \oint A \ ds \dots\dots\dots\dots\dots\dots\dots\dots \ (4.28)$$

Here, the line integral is performed outside the solenoid. This equation comes from the single-valuedness condition of the wave function

$$\hbar \oint \text{grad} \ S \ ds = mh. \ \dots\dots\dots\dots\dots\dots\dots\dots\dots \ (4.29)$$

The AB effect is produced from the second term in equation (4.28). Bocchieri, Loinger and Siragusa [79] stressed again that the second term in equation (4.28) vanishes with the non-Stokesian vector potential (see equation (4.15)).

Jánossy [98] had already pointed out in 1970 that equation (4.28) should be added to equation (4.17) to make it equivalent to the Schrödinger equation. However, he also claimed that the AB effect can be explained in terms of field strengths, and that there is thus no need to give a special role to the vector potential. His physical picture of the AB effect is as follows.

An interference pattern is obtained from an arrangement without a solenoid, as shown in Fig. 4.8(a).

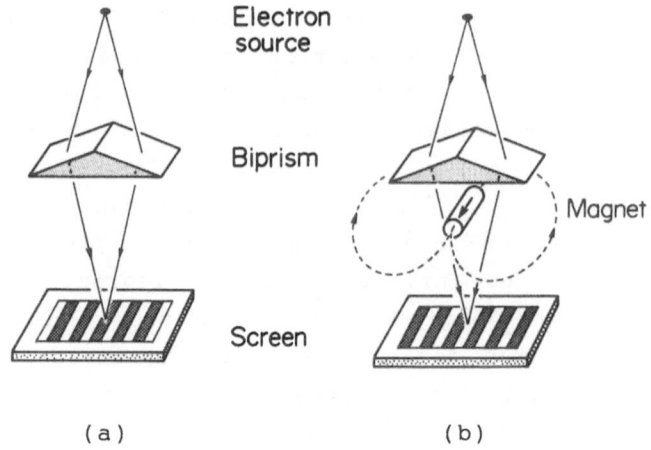

Fig. 4.8 Different boundary conditions for two cases:
(a) Φ = 0 and (b) Φ ≠ 0.

In this stationary state, the distribution ρ on the screen character-
izes the interference pattern. When a solenoid (magnet) is employed,
the magnetic field upsets the "stationarity" of the configuration and
works to initiate a new state (see Fig. 4.8(b)). Although only a
small magnetic field interacts with the electron wave, the magnetic
flux passing through the solenoid must return somewhere. Therefore,
the wave is surrounded by magnetic lines of force. This change in
boundary conditions is of great importance for determining the new
stationary state, and the effect of the solenoid is felt through the
boundary condition.

Casati and Guarneri [99] also asserted in 1979 from a hydrodynamical standpoint that vector potentials have no special
significance. The observed AB effect can be considered a consequence
of local interaction between the magnetic field and the electron
wave, however small their overlap is.

In 1984, Wódkiewicz [100] calculated the bound-state AB effect
in an explicit gauge-independent manner, using the hydrodynamical
formulation. According to his view, such a formulation has often
been used as an argument both for and against the AB effect because a
concrete calculation had never been completely carried out in a gauge-
independent manner. The calculation, here, is rather simple: a non-
local equation (4.28) can be transformed to a new form, which
contains only the field strength using the Stokes' theorem. That is,

$$\int (M \text{ rot } \mathbf{v} - e\mathbf{B}) \ \mathbf{dS} \ = mh \dots\dots\dots\dots\dots\dots\dots \quad (4.30)$$

When this equation is applied to an electron orbiting around a solenoid as shown in Fig. 4.1, the integration leads, because of cylindrical symmetry, to

$$2 \pi a M v_\theta - e \Phi = mh. \ \dots\dots\dots\dots\dots\dots\dots \quad (4.31)$$

Here, a and v_θ are the radius of the electron trajectory and the rotational component of the electron velocity, respectively. Kinetic energy, $\frac{M v_\theta^2}{2}$ can then be given by $\frac{\hbar^2}{2Ma^2} (m + \frac{e\phi}{\hbar})^2$, i.e. equation (4.22), which indicates the existence of the AB effect.

Shiekh [101] also calculated the scattering cross section of the AB effect with a gauge-free formalism which he developed using the path integral method. His results confirmed the existence of the AB effect. Liang and Ding [102] asserted from a hydrodynamical formulation that no electron scattering would occur if an infinite and inaccessible solenoid were utilized.

(3) AB effect for bound-state electrons

Bohm and Hiley [85] asserted against Bocchieri and Loinger that a gauge function, $\Lambda(x)$, must always be single-valued. Otherwise, a gauge transformation will change reality. According to their interpretation, equations (4.18) and (4.21) actually describe different situations. Equation (4.18) corresponds to an electron orbiting around an axis without magnetic flux, while equation (4.21) corresponds to the case with magnetic flux. The two situations cannot be related through a valid gauge transformation.

As a means of understanding the physical nature of the AB effect, the origin of the flux-dependence of the energy eigenvalue was explained by introducing a time-varying magnetic flux to analyze the problem of a rotating electron around the flux. This was done by Weisskopf [103], Peshkin, Talmi and Tassie [25], Noerdlinger [27], Peshkin [104], Wilczek [105] and Kobe [106]. In this view, when the magnetic flux in the solenoid is turned on and increases slowly up to Φ, an electric field, $\mathbf{E}$, circulating around the solenoid is induced from Faraday's law in the manner rot $\mathbf{E} = - \frac{\partial \mathbf{B}}{\partial t}$.

Specifically,

$$E_\theta = -\frac{1}{2\pi r}\frac{d\Phi}{dt}. \quad \dotfill \quad (4.32)$$

A rotating electron receives the same torque, N, regardless of where it is, such that

$$N = \frac{e}{2\pi}\frac{d\Phi}{dt}. \quad \dotfill \quad (4.33)$$

As a result, the kinetic angular momentum changes from $m\hbar$ to $m\hbar$ + $\frac{e\Phi}{2}$, and the energy from $\frac{\hbar^2 m^2}{2Ma^2}$ to $\frac{\hbar^2}{2Ma^2}(m + \frac{e\Phi}{h})^2$.

In 1981, Goldin, Menikoff and Sharp [107] proposed a history-dependent interpretation of the AB effect using representations of a local current algebra in a non-simply connected space. According to their interpretation, when the region of the magnetic flux is first penetrated, the AB effect can be observed even if an infinite potential barrier is afterwards introduced. Subsequent changes in the magnetic flux produce no observable effects. Consequently the vector potential plays no role in the AB effect.

In response to this assertion, Bocchieri and Loinger [108] made the following comment. The AB effect must not be confused with the Lorentz-force effect which is produced when there is penetration. This effect decreases as the penetration diminishes until the situation becomes identical to that for zero-flux.

In 1983, Frolov and Skarzhinsky [109] investigated the turning-on procedure for both scattering and bound-state AB effects. They stressed that, in addition to an effect which has a classical counterpart in the induced electric field during the turning-on process, there is also a quantum-mechanical effect connected with appearance of vector potentials in the scattering case.

Bocchieri and Loinger [110] disputed these results also. They stated the work was only a calculation of the effect of an electric field on a wave packet and had nothing to do with the AB effect.

In 1984, Roy and Singh [111] investigated the AB effect for a time-dependent magnetic flux. They proposed an experimental test to clarify speculation that the AB effect may be due to induced electric fields. If the effect truly was due to induced electric fields, they felt the interference pattern would depend not only on the present value of flux, but also on whether electrons were influenced by the electric field resulting from establishment of the flux. That is, the flux's history would be crucial. They hypothesized that if it

was truly electric field-dependent, the AB phase shift would not
arise when magnetic flux was established before the electron beam was
switched on. Such an experimental test would, Roy and Singh stated,
demonstrate   the   physical   relevance   of   non-single-valued   wave
functions.

(4) AB scattering

The scattering problem of the AB effect was further investigated by Peshkin, Talmi and Tassie [25] to elucidate some
difficulties in the physical interpretation of the results obtained
by Aharonov and Bohm [1]. Problem areas included the absence of
return flux the distortion of the incident wave at infinity, and the
singularity at the solenoid axis.

These problems were avoided by considering a case where an
incident wave is scattered by two solenoids containing opposite
magnetic fluxes, as shown in Fig. 4.9.

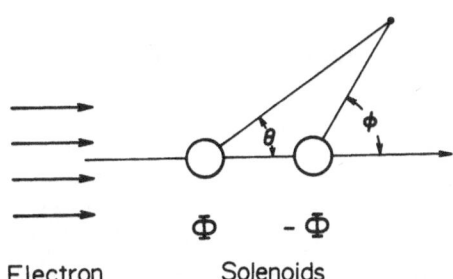

Fig. 4.9 Electron scattering by two solenoids.

The vector potential for this arrangement is

$$\mathbf{A} = \frac{\hbar}{e} \ \text{grad} \left[\alpha (\ \theta - \phi \ ) \right], \quad \dots\dots\dots\dots\dots\dots\dots \quad (4.34)$$

where θ and ϕ are defined in Fig. 4.9. $\theta - \phi$ is made single-valued
by a cut along the line connecting the solenoids. Peshkin et al.
concluded from considerations for the two cases $\alpha = 0$ and $\alpha = \frac{1}{2}$
that an inaccessible magnetic field alters the scattering of
electrons even when the return flux is taken into consideration.

The problem of return flux was later investigated in more
detail for a toroidal solenoid by Lyuboshits and Smorodinskii [112].
It was also looked at for two opposite solenoids situated
perpendicular to the incident wave direction by Olariu and Popescu
[15].

In 1962, the solution to the problem of AB scattering that had been obtained by Aharonov and Bohm [1] was criticized by Feinberg [30]. He stated that Aharonov and Bohm's solution was, strictly speaking, incorrect. The incident wave is not a plane wave, they said, but a plane wave multiplied by a phase factor, exp (-iα θ), as shown in equation (2.7).

Therefore, the wave function changes when the scattering angle, θ, is changed by 2π. Nevertheless, Feinberg did admit that essential results, such as the existence of scattering in the case of non-integer $\frac{e\phi}{h}$ (= α) and the infinite total cross section, were consistent with Aharonov and Bohm's description. He solved equation (2.6) for a case where the value of α was close to an integer, which is enough to settle the question as to whether or not scattering exists.

The wave function and cross section obtained by Feinberg are different from those of Aharonov and Bohm (see equations (2.8) and (2.9)) and can be described as

$$\Psi \rightarrow e^{-ikx} - \frac{\pi}{2} \alpha_n \tan(\frac{\theta}{2}) J_o(ka) H_o^{(1)}(kr), \quad \ldots\ldots\ldots\ldots \quad (4.35)$$

and

$$\sigma = \frac{\pi \alpha_n^2}{2k} \tan^2 (\frac{\theta}{2}). \quad \ldots\ldots\ldots\ldots\ldots\ldots\ldots\ldots\ldots \quad (4.36)$$

Here, α_n is the difference between α and the nearest integer n: $\alpha_n = \alpha - n$. Thus, there must be scattering.

In 1965, Kretzschmar [113] studied AB scattering by means of a time-dependent Green's function. His results fully supported those of Aharonov and Bohm. The difference between the Aharonov-Bohm and Feinberg cross-sections (equations (2.9) and 4.36)) was attributed by Corinaldesi and Rafeli [114] in 1978 to use of the Born approximation to derive equation (4.35).

In 1980, Berry [115] also approached AB scattering by avoiding any multi-valued wave function which cannot correctly represent the quantum state. His procedure was to decompose the incident wave function, exp(-ikx), which is single-valued, into an infinite number of "whirling waves", each of which is multi-valued. He then multiplied each whirling wave by a magnetic phase factor, exp (-i$\frac{e}{h}$ $\oint$**Ads**). Finally, he summed the phase-shifted whirling waves to get the single-valued wave function. The resultant wave function was the exact solution originally obtained by Aharonov and Bohm [1].

In 1983, Bawin and Burnel [116] studied AB scattering using a vector potential which vanishes wherever the magnetic field vanishes. This multi-valued vector potential requires a cut in space and a fiber bundle description. Such a description, together with gauge invariance, implied the existence of AB scattering and allowed the rejection of Henneberger's solution [80].

A critical overview of AB scattering, together with detailed discussions, was reported by Ruijsenaars [117] in 1983 and by Olariu and Popescu [15] in 1985. All authors concluded that the results obtained by Aharonov and Bohm were essentially correct. In 1984, Aharonov et al. [118] calculated the scattering of an electron using an infinite solenoid of finite radius for cases both of penetrable and impenetrable solenoids. The cross sections for both cases lead to the AB cross section, within the limits of a zero solenoid radius. They proceeded to explain how the first Born approximation failed to calculate the cross section. Nagel [119] then confirmed the breakdown of the Born approximation with respect to AB scattering within the limit of a zero solenoid radius. Within the limit, he stated, all orders of the Born series have to be included. Brown [120] also confirmed the validity of the AB and Born approximate solutions, which had previously been noted by Kawamura [121] in connection with an analogous problem involving the interaction of conduction electrons with crystal dislocations. In his paper, Kawamura showed how an AB effect-like phenomenon resulted from an elastic strain field in crystal. In addition to the light this shed on realities of scattering, he also pointed out how this fact might help explain the AB effect.

4.3.3 Discussions on the Validity of Early Experiments

Bocchieri and Loinger [3] expressed doubt about the existence of the AB effect not only in theoretical terms, as described in the preceding section, but also from an empirical viewpoint [77]. They concluded that although Aharonov and Bohm [1] investigated a case where an electron was never subject to a Lorentz force, the interference experiments performed up to then must have been affected by leakage fields from solenoids or whiskers.

Bocchieri and Loinger asserted that Chambers' tilted fringes [9] (see Fig. 3.1) could be fully explained by a leakage magnetic field from the whisker. The experiment by Boersch et al. [12] was then discussed in detail, and the detected fringe shift was attributed to interaction of an electron with magnetic fields.

The experimental arrangement by Boersch et al. is shown in Fig. 4.10(a).

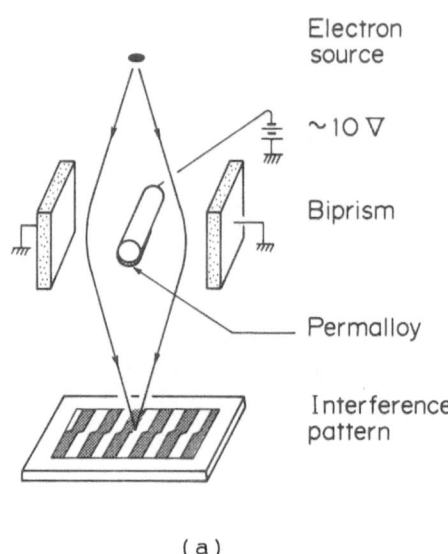

(a)

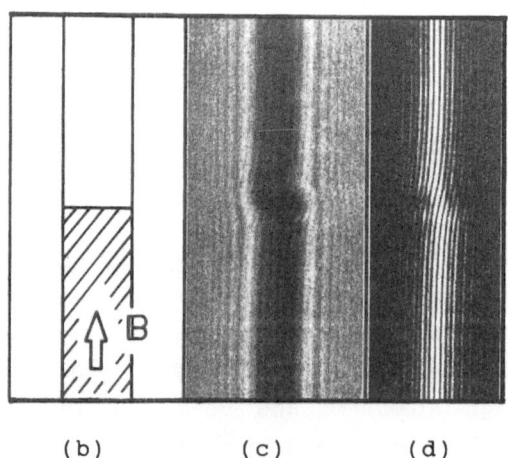

(b)   (c)   (d)

Fig. 4.10   Experiment by Boersch et al.: (a) Electron-optical system, (b) biprism filament, (c) defocused filament image, and (d) biprism interference pattern.

An interference pattern is observed with an electron biprism, the central filament of which is covered with Permalloy for only half its length and on only the lower surface (Fig. 4.10(b)). The filament shadow (Fig.4.10(c)) is displaced at the transition region due to stray fields. Meanwhile, biprism interference fringes are shifted in the opposite direction at the transition region (Fig. 4.10(d)).

These results were interpreted by Bocchieri, Loinger and Siragusa [78] as being due to the Lorentz force alone. They asserted that the action of the Lorentz force on the electron takes effect not only at the transition region, but also along the whole filament region covered with Permalloy. Judging from the shifted shadow image of the filament, the electron must, they surmised, have penetrated into the Permalloy not only at the transition region, but also along the whole filament. For an experiment to be valid, they said it would need to be carried out in the complete absence of the Lorentz force.

With regard to the rather comprehensive experiment by Möllenstedt and Bayh [18] (see Fig. 3.2), Bocchieri et al. [77] asserted that there must be a significant magnetic field in the space between two consecutive turns of the solenoid. This field, which has a non-oscillating component as well as one that is spatially oscillating, is approximately parallel to the solenoid axis. An electron entering this field would feel a Lorentz force. If the electron wave did not penetrate into the solenoid, the fringes would not be shifted; however, even a small penetration would be sufficient to generate an appreciable fringe shift.

Bohm and Hiley [85] replied to Bocchieri et al. [77]. In the Möllenstedt and Bayh [18] experiment, they said the electron beams are free from magnetic fields. This is because the solenoid was placed in the enlarged shadow of the solenoid, and also because the fringing fields from both ends of the solenoid were confined to a high mu-metal strip.

Boersch et al. [122] replied to the criticisms of Bocchieri, Loinger and Siragusa [77] against their experiment. Although they admitted that displacement of biprism fringes at the transition region could be due to the leakage field, their envelope returned to the unshifted position far from the transition region (see Fig.

10(d)). This fact cannot, they emphasized, be explained by the Lorentz force.

Bocchieri and Loinger [123] again disputed these interpretations. Since the magnetic fields do not diverge, they said the returning field outside the filament cannot be zero. The directions of magnetic fields inside and outside the Permalloy are opposite. They felt that this explains why the biprism fringes and their envelope are displaced in opposite directions at the transition region. An unobjectionable experiment, they proposed, should employ an impenetrable toroidal magnetic field.

In 1980, Roy [124] asserted using the formulation of DeWitt [28] and Belifante [29] that physical effects depend on accessible fields only, and that no effect due to inaccessible fields can exist. The magnetic fields leaking from both ends of a finite solenoid in the previously reported experiments could not be neglected. Consequently, the experimental results could not be interpreted as being caused by inaccessible magnetic fields. He stressed that a solenoid of finite length, even if surrounded by an impenetrable cylinder, yields a potential with asymptotic behavior as to exclude effects of inaccessible fields.

Roy's assertion was countered by Klein [125], Lipkin [126], Greenberger [127], and Peshkin [128]. Klein [125], as Roy, discussed the AB effect in the framework of DeWitt's line-dependent gauge. He concluded that previously performed experiments could be considered to show the reality of the AB effect even if the results were determined solely by the accessible field.

Lipkin [126] said that he saw Roy's assertion as meaning that the AB effect can be attributed to the fringing field from a finite solenoid, and not to the flux in the interior. It is true, Lipkin said, that measurement of the fringing field can always give information regarding the interior. But, he asserted, it seems absurd to attribute the observed AB effect to a fringing field which the most part of the electron wave does not experience. He felt that toroidal geometry provides a clear counter example to Roy's basic argument, for it is a finite system where the magnetic flux inside cannot be determined by knowledge of the field strengths outside.

Greenberger [127] gave an example of Mach-Zehnder type interferometers which provides good insight into why the AB effect must exist in quantum theory. He then argued that the main conclusion of Bocchieri, Loinger and Siragusa is incorrect.

In 1983, Home and Sengupta [78] supported Roy's assertion that the experimentally observed effects were due to leakage flux from the finite solenoid or whisker. According to their explanation, a loop integral of the vector potential **A** around a finite solenoid, which determines the electron phase shift, can be reduced to a surface integral of the magnetic field **B** just outside one end of the solenoid. Consequently, the phase shift can be attributed to local fields in the accessible region. Only a magnetic field trapped in a torus such as that proposed by Kuper [129], they thought, could corroborate the AB effect conclusively, because such a solenoid has no ends. At this point, Babiker and Loudon [130] claimed that this region of space close to one end of the solenoid was not sampled by the electron beams in the usual experimental setup. Therefore, the assertion by Home and Sengupta, they felt, was not to the point.

4.3.4 Various Aspects of AB Effect

The AB effect is deeply rooted in quantum mechanics and has frequently been discussed in relation to fundamental problems such as the single-valuedness of the wave function, the quantization of angular momentum, topology, flux quantization, magnetic monopoles, and non-locality. Such problems relate not only to the AB effect, but to different aspects of fundamental physics.

(1) AB effect and angular momentum

Peshkin, Talmi, and Tassie [25] discussed the AB effect for bound-state electrons as an attempt to clarify both the theoretical aspects of scattering and the experimental aspects of return flux [128]. The energy eigenvalues, they said, depend on magnetic flux in the region inaccessible to the electron, which can be described as the bound-state AB effect.

In 1981, Peshkin [104] further investigated the physical meaning of the flux dependence of energy eigenvalues using a slightly different arrangement (see Fig. 4.11).

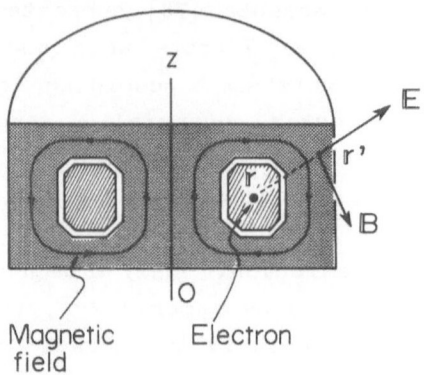

Fig. 4.11   Aharonov-Bohm effect for bound-state electrons.

There, the electron was confined within the inner torus. The magnetic flux threads and circulates the torus, but the two regions do not overlap. The only interaction occurs where the electric field, **E**, of the electron penetrates into the region of the circulating magnetic field, **B**, which produces electromagnetic angular momentum, **L**.

The z-component, L_z, can be derived from

$$\mathbf{L} = \frac{1}{4\pi} \int \mathbf{r}' \times \left[\mathbf{E}(\mathbf{r}, \mathbf{r}') \times \mathbf{B}(\mathbf{r}') \right] \, d\mathbf{r}' \dots\dots (4.37)$$

to be $-\frac{e\Phi}{2\pi}$. The total angular momentum component, J_z, is given by

$$J_z = K_z - \frac{e\Phi}{2\pi}, \dots\dots\dots\dots\dots\dots\dots\dots\dots\dots\dots (4.38)$$

where K_z is the z-component of kinetic angular momentum $\mathbf{K} = \mathbf{r} \times (\mathbf{P} + e\mathbf{A})$. Thus, the z-component of canonical angular momentum, $\mathbf{r} \times \mathbf{P}$, is equal to J_z in the cylindrical gauge. Although J_z must be quantized, it is the kinetic angular momentum **K** that is contained in the Schrödinger equation.

Peshkin's conclusion was that the AB effect is a direct consequence of the flux dependence of kinetic angular momentum or of the quantization of the canonical angular momentum. Attempts to remove the AB effect must, Peshkin said, involve a drastic change in our understanding of the quantization of angular momentum.

Bocchieri and Loinger [108] asserted in 1981 that Peshkin's idea was inconclusive and that no definite conclusion about the quantization of angular momentum can be reached by merely considering electromagnetic theory. They purported that both J_z and L_z could be quantized with equal validity.

In 1982, a new controversy arose. Questions were posed as to the fundamental meaning of angular momentum. Wilczek [105] presented a novel theoretical idea about an electron-bound flux line. He asserted that, in a system where an electron orbits around an impenetrable solenoid, eigenvalues of the rotation generator can have a continuous value, such as $m\hbar + \frac{e\phi}{2\pi}$, rather than $m\hbar$. This produces, he said, unusual results due to changes in the statistics.

It is commonly taught to students of quantum mechanics that a small rotation (θ) of a system around the z-axis produces a new state described by the original wave function multiplied by an operator, $\exp(iJ_z\theta/\hbar)$, where J_z is the z-component of the angular momentum. Therefore, in an axial-symmetric case, the angular momentum around the axis is conserved and is quantized as $m\hbar$.

In the present case, the problem is whether or not the kinetic angular momentum is really given by the rotation generator and thus open to quantization as proposed by Wilczek. His assertion was clarified by Goldhaber [131], Lipkin and Peshkin [132], Jackiw and Redlich [133], and Silverman [134]. They stated that the rotation operator corresponds to canonical angular momentum. When the magnetic flux is gradually turned on, the kinetic angular momentum, K_z, becomes $m\hbar + \frac{e\phi}{2\pi}$. However, the canonical angular momentum, J_z, is conserved due to the contribution of L_z, which is the electromagnetic angular momentum distributed over the outside of the solenoid, as was discussed just a few paragraphs above (see equation (4.38)).

An experimental test for the half integer eigenvalues of the angular momentum spectra was proposed by Silverman [135]. A similar experiment was also proposed by Horváthy [136], and further studies on this subject were reported by Liang [137] and Morandi [138]. Morandi showed from an exact quantum-mechanical calculation of the path integral that there was no room for anomalous quantization of the kinetic angular momentum even in the presence of magnetic flux. He asserted that an electron interference experiment with a toroidal magnet, which had already been made by Tonomura et al. [139], could provide an answer, at least in part, to the questions posed by Silverman [135]. He pointed out that the configuration spaces for the two experiments using toroidal and double magnets had similar topologies.

(2)   The AB effect and topology

    The topological nature of the AB effect has been emphasized through formulations of electromagnetism in terms of non-integrable phase factors.   Such a formulation is described in purely geometrical terms by fiber-boundle theory [2,51].   In the context of this theory, a non-integrable phase factor corresponds to parallel transport in a fiber bundle.   To clarify this in electromagnetic terms, the electromagnetic field and vector potential respectively correspond to the curvature of the bundle and a principal connection.

    The AB effect corresponds to a purely mathematical theorem in a geometrical arrangement, as shown in Fig. 4.12.

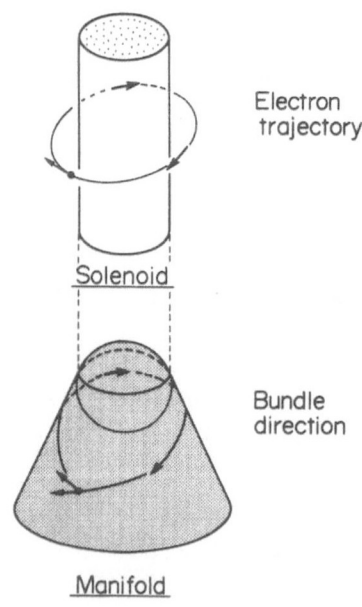

Fig. 4.12   Fiber-bundle description of AB effect.

The space around a solenoid corresponds to a truncated conical surface capped with a sphere. The curvature is nonzero only on the spherical surface, which corresponds to the interior of the solenoid. Although the bundle direction seems at first to be unchanged by parallel transport on the conical surface where the curvature vanishes (**B** = 0), a difference in bundle direction is evident after one turn around the conical surface. This precisely simulates the AB effect.

The topological characteristics of the AB effect have also been discussed using path integral formalism which was originally devised by Feynman [57]. In 1971, Schulman [140] became the first person to investigate the AB effect as an example of approximate topologies by utilizing this formalism. He gave the following interpretation of the AB effect after stating that the physical consequences of the AB effect could be obtained without use of electromagnetic potentials: "I exclude the electron from the solenoid, create an ambiguity, and look to physical quantities (magnetic flux) to resolve this ambiguity. Aharonov and Bohm also exclude the electron from the solenoid but retain a vector potential which remembers what is going on inside. These are just two different ways to do quantum mechanics." He described the problem of multiple solenoids as one of great potential interest in the context of this formalism.

Inomata and Singh [141] proposed a method in 1978 to evaluate path integrals under a periodic constraint. The AB effect was then investigated using this formalism (Bernido and Inomata [142], [143]). In addition to the flux-dependent phase shift, a pure topological shift was predicted to exist which depends on the winding number (see Fig. 4.13). This result was asserted to contradict Schulman's conclusion [140] that the AB effect can be confirmed mathematically without use of electromagnetic potentials.

The winding number dependence of the AB effect was experimentally demonstrated using a superconducting ring by Deaver and Donaldson [144] in 1982. The experimental setup is shown in Fig. 4.14.

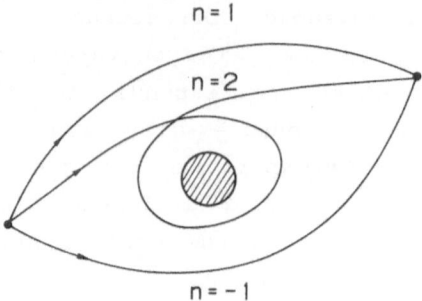

Fig. 4.13 Winding number.

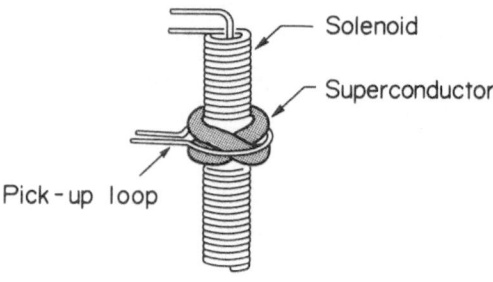

Fig. 4.14 Deaver and Donaldson experiment.

A two-turn superconductor ring is looped around a solenoid. Change in flux was detected as an induced current through a pick-up coil when the temperature crossed a superconducting critical temperature. Results indicated that the magnetic flux is quantized in h/4e units. This experiment was suggested by Yang [144] to be equivalent to one in which a one-turn superconducting ring is threaded twice by a magnetic flux, as shown in Fig. 4.15.

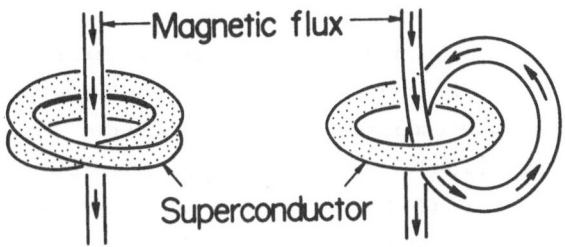

Fig. 4.15   Equivalence of two physical situations
            (Yang [144]).

     This interpretation was disputed by Gerry and Singh [145]   as
well as by Inomata [146].   The numbers of ring turns, they state,
are not totally independent of each other.   Even for a single ring,
all possible values of the winding number can contribute.   A method
was suggested for detecting the effect of winding numbers.

     In 1979, Kobe [6] calculated a double-slit diffraction pattern
using the path integral method for different configurations of a
magnetic field.   He did so to clarify the difference between the
effects of the magnetic field and the effect of the vector potential
only.   The force on the electron, he stated, causes a shift in the
whole diffraction pattern, whereas the potential only shifts the
fringes in the unchanged envelope.   He also showed that results
approached what could be expected from a classical interpretation
for both $h \to 0$ and $\lambda \to 0$.   Thus he concluded that the AB effect is
purely quantum-mechanical.

     In 1984, Morandi and Menossi [147] investigated (also using the
path integral method) the AB effect in the context of a simplified
two-dimensional model.   They clarified the periodic shift of
interference fringes in response to a magnetic flux. The calculated
wave function was confirmed by comparison with an exact expression
of the AB wave function that had been obtained by Berry [115].

     In 1986, Ohnuki [148] examined the AB effect, once again using
the path integral technique.   He concluded that the single-
valuedness of the wave function is a necessary consequence of

quantum mechanics, thus leading to the explanation described as the AB effect. Inomata [149] proposed a double-solenoid interference experiment to observe both the AB interference with a pair of higher winding numbers and its winding number dependence.

In 1984, Berry [150] predicted the existence of a general topological phase factor (Berry's angle) which arises when a system is altered adiabatically and then returned to its original state. Simon [151] considered the phase factor to arise, in mathematical terms, from parallel transport along a closed path in the parameter space. This factor can be isolated by observing the interference between perturbed and unperturbed systems. According to this interpretation, the AB effect can be understood as a special case for the geometrical phase factor. Some other predicted interference effects were detected by Tomita and Chiao [152], and Delacretaz et al. [153]. Further theoretical investigations of Berry's phase were carried out by Anandan and Stodolsky [154], and Aharonov and Anandan [155]. The latter pair of authors predicted a more general geometric phase factor for cyclic evolutions which they asserted, would extend beyond the scope of Berry's phase.

(3) AB effect and flux quantization

Back in 1961, magnetic flux trapped in a hollow superconducting cylinder was discovered to be quantized in h/2e units by Deaver and Fairbank [41], and Doll and Näbauer [156]. They precisely measured the torque exerted on the magnetic moment of the cylinder. Although a possible flux quantization had already been suggested by London [157] and Onsager [158], Byers and Yang [159] concluded that this phenomenon involves no new physical principle and merely indicates the pairing (Cooper pair) of the electrons in the superconductor. Further investigation into the persistent current phenomenon was made by Peshkin [160].

Byers and Yang [159], as well as Peshkin [104] pointed out that flux quantization is based on the same principle as that of the AB effect; that is, physical effects are determined by inaccessible magnetic flux. Here it would be good to step aside and take a deeper look at the principle behind flux quantization.

If a wave function for Cooper pairs is assumed to exist and also to be describable with the Schrödinger equation, then kinetic momentum 2M**v** is given (see equation (4.17)) by

$$2M\mathbf{v} = \hbar \ \text{grad} \ S + 2e\mathbf{A}. \quad \dots\dots\dots\dots\dots\dots \quad (4.39)$$

This equation can be applied to the arrangement shown in Fig. 4.16, where magnetic flux is trapped in a hollow superconducting cylinder.

The integration of equation (4.39) along a loop within the superconductor leads to

$$2M \oint \mathbf{v} \ ds = \hbar \oint \ \text{grad} \ S \ ds + 2e \oint \mathbf{A} \ ds. \quad \dots\dots\dots\dots \quad (4.40)$$

If the wave function is assumed to be single-valued, then

$$\oint \left(\frac{M\mathbf{v}}{e} - \mathbf{A} \right) ds = \frac{\hbar}{2e} \oint \text{grad} \ S \ ds = \frac{nh}{2e} , \quad \dots\dots\dots\dots \quad (4.41)$$

where n is an integer. This equation means that the magnetic flux $\Phi (= \oint \mathbf{A} \ ds)$ is not always quantized in h/2e units.

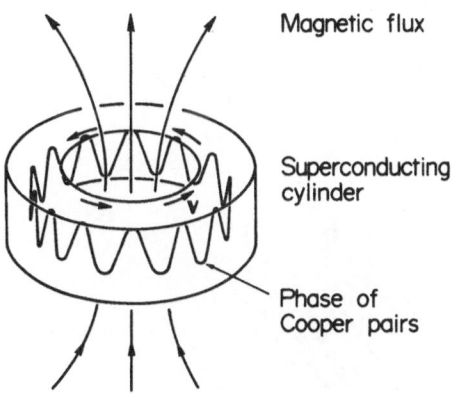

Fig. 4.16 Magnetic flux quantization.

Two conclusions can be derived using this equation.

° When the thickness of the hollow cylinder is much larger than the penetration depth, the magnetic flux is quantized in h/2e units. This is because the supercurrent is localized only in the inner surface region and vanishes inside the superconductor (**v**=0).

° When the thickness, d, of the hollow cylinder of radius r is smaller than the penetration depth, λ , the value of the flux quantum is not given by h/2e. According to Bardeen [161], the trapped flux is quantized in flux units like this:

$$\frac{h}{2e} \ (1 + \frac{2\lambda^2}{rd})^{-1} \quad \dots\dots\dots\dots\dots\dots\dots\dots\dots\dots\dots \ (4.42)$$

The flux quantum, h/2e, produces a phase shift of π between two enclosing electron beams (see equation (2.5)). This phase shift was experimentally detected by Wahl [43] and Lischke [162]. Lischke located a hollow superconducting cylinder, which traps magnetic flux, in place of the central filament of an electron biprism. Wahl did so in the shadow of the filament (see Fig. 4.17).

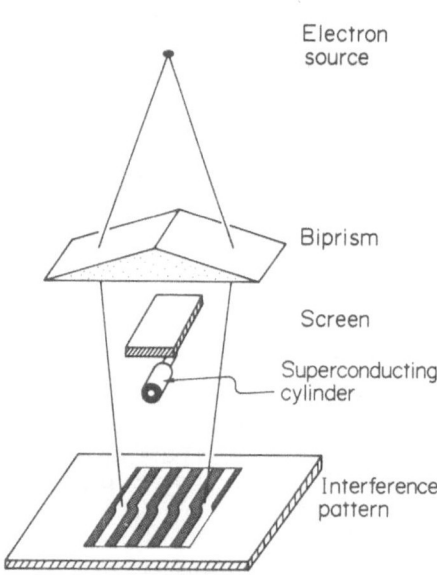

Fig. 4.17 Electron interference experiment to detect a single flux quantum.

The value of the trapped flux can be measured as a shift of the biprism interference fringes by utilizing knowledge of the AB effect.

Controversy continued to rage as to the existence or nonexistence of the AB effect. At this juncture an unconventional assertion regarding magnetic flux quantization was made by Costa de Beauregard and Vigoureux [163]. They predicted that the magnetic flux inside an "autistic" (infinitely long, or toroidal) magnet is quantized in h/2e units due to the presence of an evanescent electron wave outside the magnet. A factor of 1/2 does not come from the electron pairing as in the case of superconductivity, but from the electron spin, $\hbar/2$. They asserted that the flux quantization demonstrated by Deaver and Fairbank, and Doll and Näbauer, might be the only way to realize an ideal "autistic" magnet.

An attempt at deducing electromagnetic phenomena was made by Jehle[164]. He regarded quantized magnetic flux as a primary entity rather than a consequence of conditions imposed on the supercurrent. The electric field, he said, is produced by the motion of the flux. Furthermore, he considered the properties of elementary particles to be determined by the topological structure of their magnetic fields. Post [165] regarded the possibility of the quantization of electroflux, $\int \varphi(t)dt$, occuring in h/2e units as a necessary companion of magnetic flux quantization.

(4) AB effect and magnetic monopole

The AB effect can be viewed as having some relationship to a magnetic monopole. That is, the quantization condition of a magnetic monopole cannot be explained without accepting the existence of the AB effect (see Lipkin and Peshkin [132], and Kunstatter [166]). Furthermore, both the AB effect and magnetic monopoles can be regarded as global gauge-invariant manifestations of electromagnetism ([2],[167]).

At this point, let's turn to look at the interaction of an electron wave with a transparent magnet, an infinite solenoid and a semi-infinite solenoid. The semi-infinite solenoid can be virtually regarded as a monopole. The idea is that such discussions will help in understanding the relation between the electron phase and vector potential.

The first case is shown in Fig. 4.18.

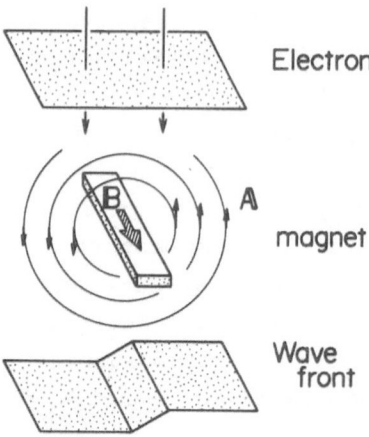

Fig. 4.18   Interaction of electron wave with
             magnetic fields: magnetic thin film.

Parallel electrons incident on the transparent magnet are deflected
by the magnetic field inside it as they pass through it.   If the
wave front is assumed to be perpendicular to electron trajectories,
the incident parallel electrons correspond to a plane wave.   The
wave front transmitted through the magnet is a tilted plane with the
right side up.   To summarize, the incident wave front can be said to
be tilted on a rotation axis determined by a magnetic line of force.

    If the wave front is assumed to be continuous, the resultant
wave front level differs on both sides of the magnet as illustrated
in the figure.   Although the wave front there does not intersect any
magnetic line of force, the reason for the wave front displacement
can easily be understood as resulting from the vector potential.   The
amount of displacement calculated from the deflected angle is equal
to the amount calculated from the AB phase shift.   The amount of
displacement does not depend on electron energy, but only on the
magnetic flux.

    This energy independence is very important to the existence of
the AB effect, especially when discussed from an experimental point
of view.   Penetration decreases as the electron energy is lowered.
The value of the displacement, however, remains unchanged even at
the limit of zero electron energy where no penetration takes place.

    The limiting case corresponds to the AB effect with an
impenetrable solenoid shown in Fig. 4.19.

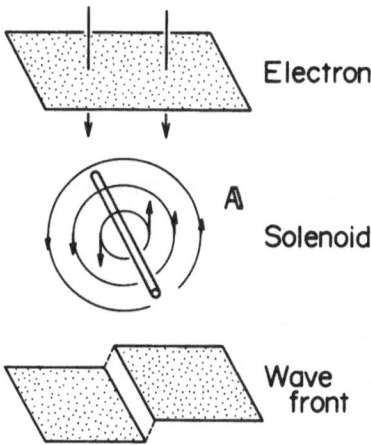

Fig. 4.19   Interaction of electron wave with
magnetic fields: infinite solenoid.

In this case, the wave front displacements on opposite sides of the
solenoid cannot be explained from wave front continuity. Thus,
it is necessary to rely on the existence of physical effects
incurred by a vector potential. Unless the total magnetic flux is
equal to nh/e (n: integer), the effect of inaccessible flux will be
found in an electron interference pattern.

What happens if an infinite solenoid is cut into two parts and
an electron wave front is incident on one of two semi-infinite
solenoids? This is the third case illustrated in Fig. 4.20. A
magnetic flux leaks outside from the end of the semi-infinite
solenoid; it looks like a magnetic monopole accompanying the
solenoid (see 't Hooft [168]).

An incident wave front is rotated on axes determined by these
radiating magnetic lines of force, and the resultant wave front
appears like a spiral staircase. Starting from a point just beside
the solenoid and moving around the monopole, the wave front gradually
rises. The position arrived at after one round, just on the opposite
side of the solenoid, is not the same as the original position in
terms of wave front level. Rather, it is on a second level of the
wave front. When we cross the solenoid, the wave front returns to
the original position due to the AB effect.

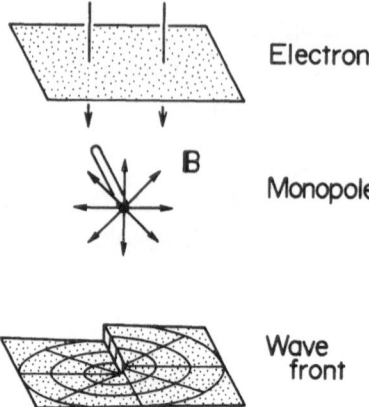

Fig. 4.20   Interaction of electron wave with magnetic
            fields: semi-infinite solenoid.

This semi-infinite solenoid cannot, in general, simulate a
magnetic monopole, because the monopole does not accompany such an
observable solenoid.   In order to make the solenoid invisible, the
flux value inside the solenoid has to be nh/e as was pointed out by
Ferrell and Hopfield [169], and Lubkin [170].   The resultant displacement of the wave front on the opposite sides of the solenoid
becomes 0 modulo 2π.   This is Dirac's quantization condition [171]
for the monopole charge.   Under this condition, a free monopole
cannot be discriminated from a semi-infinite solenoid of zero diameter, i.e. a string.

The explanation on the last page has presented the change in the
wave front in terms of an interaction with magnetic lines of force.
This change, however, can also be explained in terms of an interaction with the vector potential, **A**.   In this case, however, difficulties appear regarding a vector potential around a monopole.   First of
all, **B** = rot **A** and div **B** $\neq$ 0 imply that **A** has singularities at some
points.

Dirac's vector potential (see Goddard and Olive [172]),

$$\mathbf{A} = \frac{g(\mathbf{r} \times \mathbf{n})}{r(r - \mathbf{rn})} \; , \; \dots\dots\dots\dots\dots\dots\dots\dots\dots\dots\dots\dots\dots \quad (4.43)$$

where **n** is an arbitrary unit vector indicating the string direction, diverges along a line given by r = **rn**. This vector potential is illustrated in Fig. 4.21, together with the change in the electron wave front during passage through the potential.

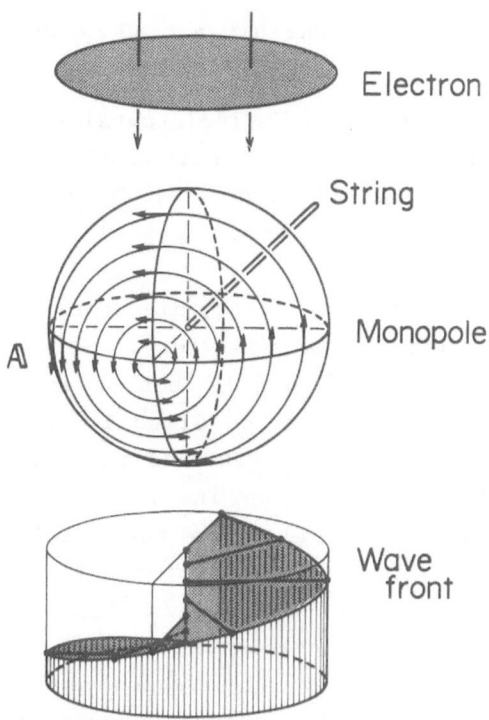

Fig. 4.21   Interaction of electron wave with
             vector potentials around monopole.

Wu and Yang [2] avoided this difficulty by dividing the space surrounding the monopole into two regions, in which different vector potentials are defined. The two potentials can be related by a gauge transformation in the overlapping regions (see Fig. 4.22), provided that Dirac's quantization condition is satisfied. Although the electron wave front transmitted through these vector potentials is discontinuous across the boundary of the two regions, this discontinuity cannot be observed since the phase shift is an integral multiple of 2π.

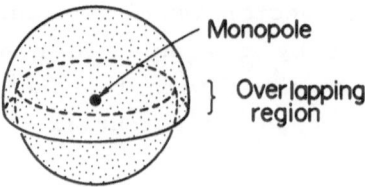

Fig. 4.22   Vector potentials around monopole
(Wu and Yang [2]).

In 1983, Roy and Singh [173] obtained a solution for an
electron wave function in the potential of a solenoid and a
monopole.   Further investigations, by Barut and Wilson [174] in
1985, looked at electron motion in a field where both a monopole
and solenoid existed.

More recently, Saffouri [175] discussed the monopole and AB
effect expressing several unusual views. He asserted, as did Barut
[176], that it is not a monopole but a string that is a real
physical entity, since an electron directly couples with the string
and not with the monopole.   In addition to demonstrating the
existence of several difficulties involved in the monopole concept,
Saffouri then showed by calculations that electrons could actually
be scattered by a string having a magnetic flux of nh/e (this
disagreed with the results obtained by Aharonov and Bohm).   This
interaction depends on the value, n, of the flux contained in the
string.   Saffouri stated that the reason why previous experiments
could not detect such interaction because they were based on
geometrical-optics.

Since gauge transformations correspond to a change in the
position of the string, he concluded that the two situations before
and after a gauge transformation were physically distinguishable.
This conclusion was, he asserted, a partial answer to the question
raised by Aharonov and Bohm [1] regarding the physical significance
of different gauges.

(5)   Non-locality of the AB effect
Aharonov [177] stressed in 1983 that the AB effect is an
example of non-local phenomena since the effect of the vector

potential in any local region is not gauge invariant and therefore not measurable. About this assertion, Yang remarked: "The AB effect is the result of a local equation of motion in the Heisenberg representation, but with non-commuting dynamical variables. Electrodynamics is not non-local in quantum mechanics. Information about the non-integrable phase factor (4.11) for all closed loops correctly describes electromagnetism."

Spasskii and Moskovskii [178] discussed the non-local properties of the AB effect together with the Hanbury Brown-Twiss effect and EPR paradox. These phenomena, they purported, can be interpreted from the classical point of view as manifestation of a kind of non-locality inherent in quantum objects. They pointed out that the AB effect can be regarded in this view as proof of action at a distance. In actuality, however, the quantum nature of the phenomenon does not allow such a conclusion. They stated that the dichotomy between action at a distance and local interaction thus ceased to be valid.

In 1984, Van Kampen [179] posed an interesting question "Can the AB effect transmit signals faster than light?" He based a hypothetical experiment on a conventional arrangement for the AB effect, except that he proposed an extremely long distance between the solenoid and screen (see Fig. 4.23).

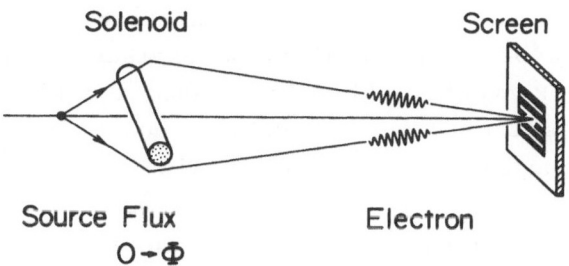

Fig. 4.23 "Can the AB effect transmit signals
faster than light?" (Van Kampen [178]).

In his scheme, the magnetic flux is first kept at $\Phi=0$, until the electron has almost reached the screen; at that moment, the flux is switched on. Then, the wave function must pick up the AB phase difference due to the single-valuedness requirement. The effect of switching-on manifests itself instantaneously to the observer at the screen, thus violating the theory of relativity. Van Kampen analyzed this as meaning that the electric AB effect affected the phase of the wave function in a non-local way as to cancel the magnetic AB effect.

In 1985, Troudet [180] discussed the experiment proposed by Van Kampen in more detail, using the path integral method. He arrived at the same conclusion as Van Kampen [179], but did so after progressing through canonical lagrangian formulation utilizing the vector potential.

4.3.5 Proposals Regarding New Experiments

The controversy regarding the AB effect, which I've described only partially up to this point, appeared as if it might continue endlessly. It is not surprising, therefore, that after 1980 a number of new experiments were proposed to serve as crucial tests for determining the existence of the AB effect.

In 1981, Kuper [129] proposed an electron diffraction experiment employing a hollow superconducting torus as shown in Fig. 4.24. A toroidal magnetic field is trapped inside the hollow of the superconductor; consequently, the total flux is quantized as nh/2e (n: the number of flux quanta). In his scheme, an electron wave is incident on the torus. The two partial waves, having passed inside and outside the torus, are superimposed on a screen to form a diffraction pattern. Since geometrical dimensions are similar to those for a zone plate, the central intensity of the diffraction pattern is bright or dark depending on whether there is an even or odd number of flux quanta. This is because the two waves receive a relative phase shift of 0 or π modulo 2π due to the AB effect; they interfere constructively or destructively at the center of the screen.

Roy [124] purported that all previous experiments can be explained as resulting from electromagnetic fields outside the solenoid. He felt that this explanation would continue to be valid, unless an ideal but unattainable geometry for an infinite solenoid were adopted. One approach he did not deal with, however, was to attempt an experiment realizing an ideal geometry via use of a toroidal magnetic field.

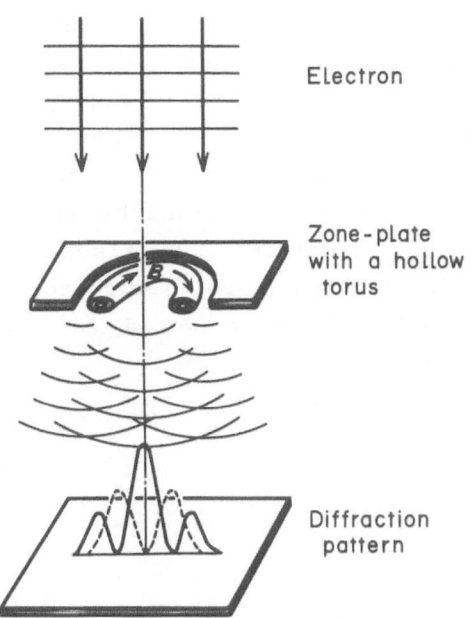

Electron

Zone-plate
with a hollow
torus

Diffraction
pattern

Fig. 4.24   Proposed experiment to confirm AB effect
(Kuper [130]).

Utilization of toroidal geometry for a solenoid has repeatedly been proposed during controversial discussions on the AB effect. Perhaps Tassie [181] was the first to actually suggest a toroidal solenoid, within which a magnetic field would be entirely contained. He went on to calculate a cross section for two cases, $\alpha = 0$ and $\alpha = 1/2$, using the vector potential given by equation (4.34)(see Fig. 4.9).

Lyuboshitz and Smorodinskii [112] in 1978 theoretically investigated the AB effect for the case of a toroidal solenoid. The total cross section of scattering, which diverges for an infinite solenoid, becomes finite for a toroidal solenoid.

Several people, such as Klein [125], Lipkin [126], Greenberger [127], Bocchieri and Loinger [123], Rothe [182], and Takabayasi [94] variously promoted this idea. Greenberger specifically posed the question as to why no toroid experiment had been performed although the geometry itself appeared quite usable. What is more, he actually carried out a neutron beam experiment employing a toroidal magnet to test the existence of the AB effect. Results were negative. Bocchieri and Loinger [123] asserted in a paper against Boersch et al. [12] that, in previous experiments, magnetic fields outside a finite solenoid were not zero. They felt that an unobjectionable experiment would need to approximate very closely the conditions of an impenetrable toroidal solenoid.

Rothe [182] admitted Roy's assertion that the vector potential outside a finite solenoid is determined by the magnetic fields in a region accessible to the electrons. He also asserted, however, that although the AB effect cannot be eliminated even for a finite solenoid a more conclusive test could be carried out using a toroidal configuration.

5.   RECENT EXPERIMENTS

As we've seen, there has been much argument over the existence of the AB effect. However, for the most part, it has been from a theoretical point of view. The reader might wonder why the problem was not empirically confirmed by some sort of experiment, early on. The answer to this question lies in the general difficulty surrounding electron interference experiments.

Since the wavelength of an electron is so short (usually 0.03Å), experimental setups have to be extremely minute, even if a coherent field-emission electron beam (see Tonomura [183]) is employed. An example of a viable experiment might involve a sample size of less than 10 μm, so that the whole sample could be fully contained in the coherent region of the illuminating electron wave. The requirement for this kind of structure made it extremely difficult to carry out experiments which might be more complete than those performed in the early 1960s. By the 1980s, though, technological progress - particularly in such areas as microlithography techniques developed for the manufacture of semiconductor devices - opened up a window of opportunity for fabrication of such tiny and complicated samples.

5.1 Experiment Using Toroidal Magnet [139], [184]

In 1982, the first experiment using a toroidal magnet was carried out by Tonomura et al. [139]. In this experiment, a newly developed technique called "electron holography" (see Tonomura [185]) was employed to quantitatively measure the leakage of magnetic flux in a microscopic region. Experimental results confirmed the existence of the AB effect, eliminating ambiguity regarding the leakage flux effect. Details follows.

5.1.1 Electron Holography

Electron holography played an important role throughout this experiment. Thus, it would be good to start by outlining this technique.

Electron holography [186] is a two-step imaging method. An interference pattern (hologram) between an object wave and a

reference wave is formed using an electron wave.   Then, an optical
image of the original object is reconstructed in three dimensions by
illumination of the hologram with a light wave (see Fig. 5.1).

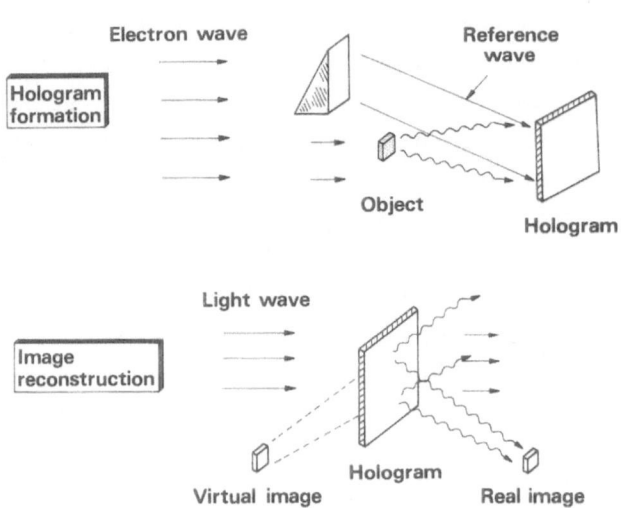

Fig. 5.1 Principles underlying electron holography.

One might wonder why such a seemingly simple imaging method is
really possible (this imaging method requires no lens, but only
utilizes interference phenomena).   Furthermore, the imaging can be
transferred between two completely different kinds of waves (the
wavelengths of electron and light waves differ by a factor of
200,000).   Although this question can be answered more exactly using
wave equations, the answer can be summarized as follows.

A lens is not the only possible device for image formation.
Since all information (intensity and phase) regarding an electron
wave is recorded in an interference pattern (except, strictly
speaking, for the sign of the phase), it is not really strange that
the image of an original object can be reproduced by holography.

The simplest example of a hologram is a Fresnel zone plate (see
Fig. 5.2), since it is the interference pattern between a plane wave
and a spherical wave scattered from a point object.

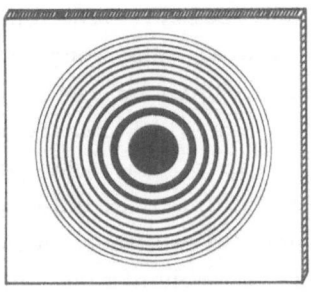

(a)

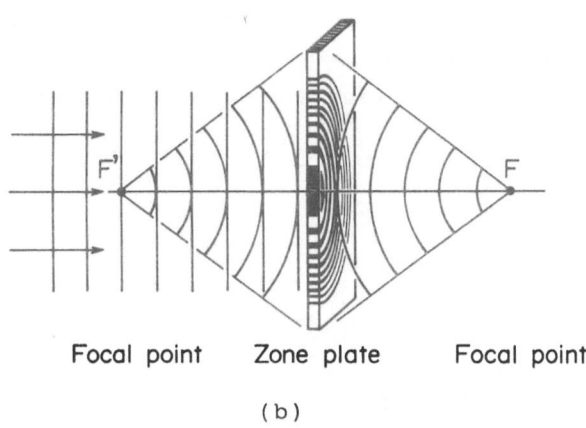

Focal point   Zone plate   Focal point

(b)

Fig. 5.2 Hologram of a point object: (a) Fresnel
          zone plate, and (b) image reconstruction.

The plate performs the same functions as both concave and convex
lenses when illuminated with coherent light. Scattered waves from
the transparent parts of the zone plate are in phase at focal points
F and F'. They form real and virtual foci.

The problem of different wavelengths is simpler to answer. An
interference pattern (hologram) is always given in units of
wavelengths, and diffraction (image reconstruction) also takes place
in units of wavelengths. Therefore, even if two waves with different wavelengths are employed for the two processes, holographic

images can be formed almost perfectly. The only exception is that scale factors, such as the lateral and longitudinal magnifications of the image, depend on the ratio between the wavelengths.

Once the electron wave front is transformed into an optical wave front via electron holography, the use of optical techniques helps to obtain new information inaccessible by means of electron microscopy. This is because there are no convenient electron counterparts for such optical components as half mirrors, concave lenses and various types of interferometers. Holographic interference microscopy, in particular, has proven effective in providing several types of new information.

With regard to magnetic samples, the contour fringes in an interference micrograph are a direct manifestation of the projected magnetic lines of force appearing in flux units of h/e. This can easily be understood by considering the contour map of the electron wave front shown in Fig. 5.3.

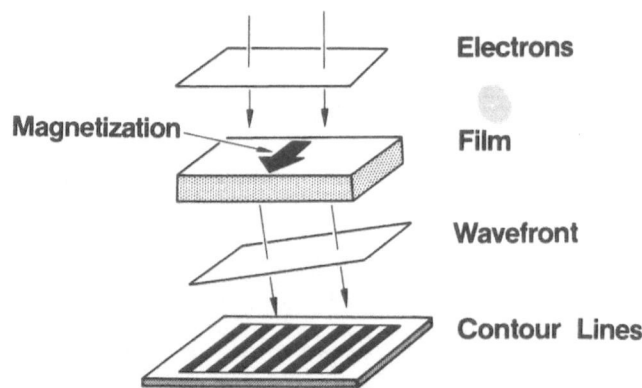

Fig. 5.3 Interaction of electron wave front
with magnetic fields.

The incident wave front is influenced by a uniform magnetic field such that the wave front is rotated around an axis determined by a magnetic line of force. The wave front level changes by a wavelength for every flux unit of h/e. This can easily be verified if electron trajectories are assumed to be perpendicular to the wave front.

This may remind you of a superconducting flux-meter SQUID. With a SQUID, small amounts of magnetic flux can be measured in flux

units of h/2e, instead of h/e, since the SQUID employs the interference phenomenon of a Cooper pair which has an electric charge of
2e. Thus, we might call electron interference microscopy, "SQUID in
a microscopic world". A concrete example is shown in Fig. 5.4,
where in-plane magnetic lines of force inside a Co fine particle are
observed. The contour fringes in the interference micrograph directly show magnetic lines of force.

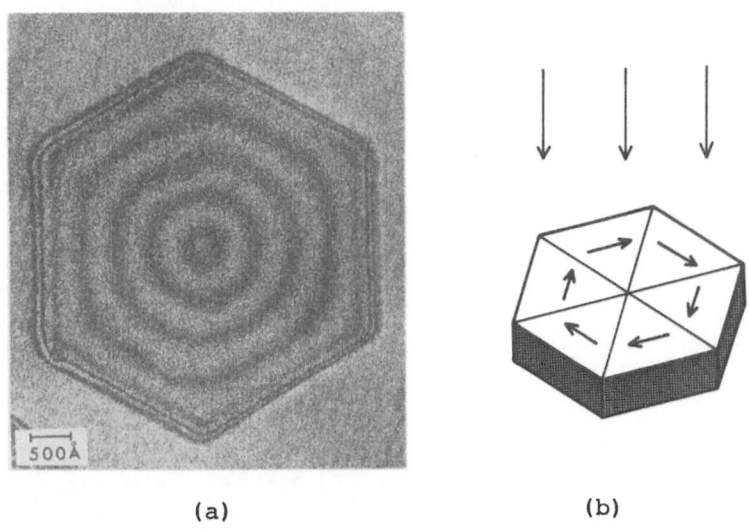

(a)   (b)

Fig. 5.4 Cobalt fine particle: (a) Interference
          micrograph, and (b) morphology.

When samples are made of a nonmagnetic substance, the contour
lines of the transmitted electron wave front indicate the thickness
contours. An example of MgO fine particles is shown in Fig. 5.5.
The phase distribution can be amplified in the optical reconstruction
stage (this method will be explained in Subsection 5.1.2). Twice
amplified interference micrographs are shown in Fig. 5.5 (c) and (d).

Reiteration of the phase amplification process makes it possible
to measure much more precisely the electron phase shift. While the
minimum phase shift previously detected with an electron biprism was
$2\pi/2$, a phase shift of $2\pi/50$ was shown to be detectable by Tonomura
et al. [187] by measuring the monatomic step height of a molybdenite
thin film (see step A in Fig. 5.6).

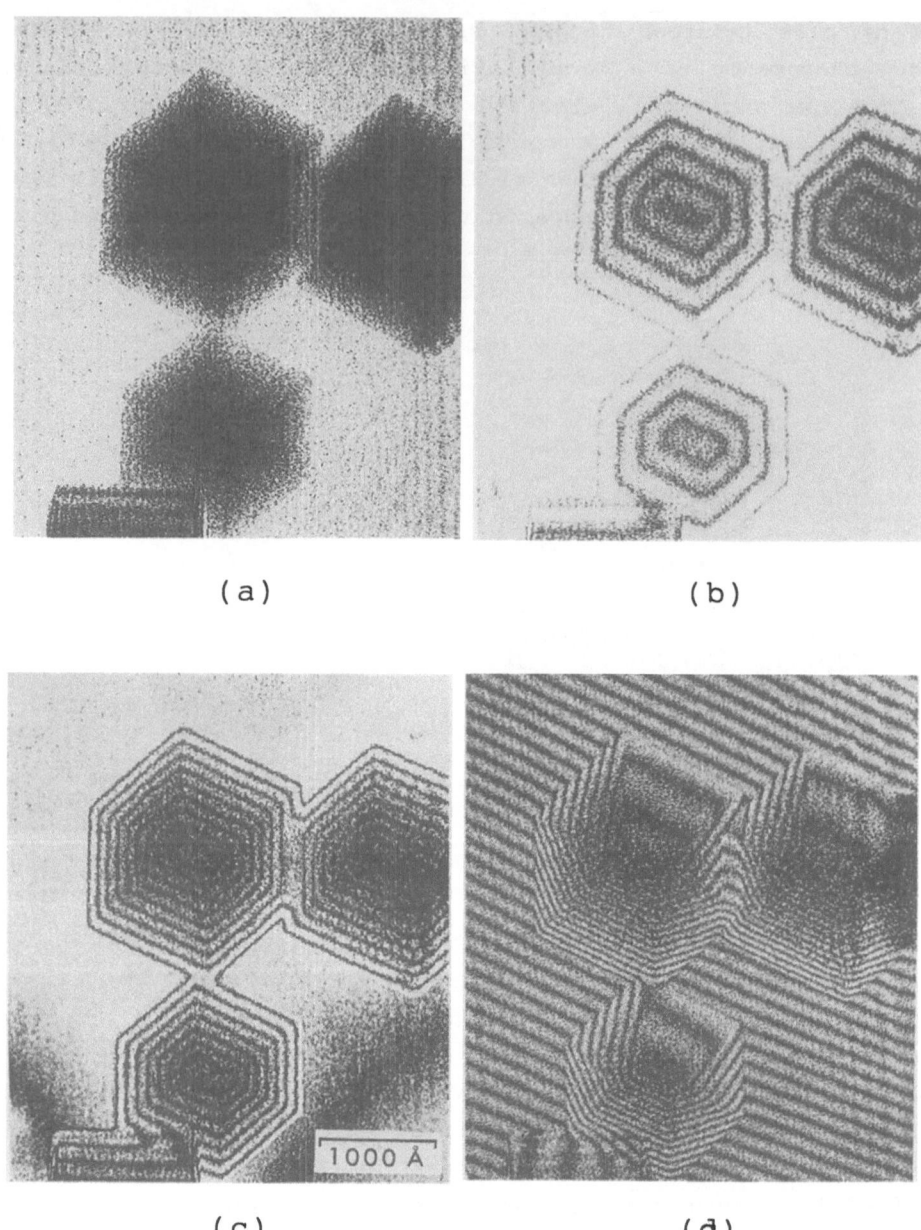

(a)

(b)

(c)

1000 Å

(d)

Fig. 5.5 Magnesium oxide fine particles: (a) Reconstructed image, (b) contour map, (c) amplified
controur map (x2), and (d) amplified interferogram (x2).

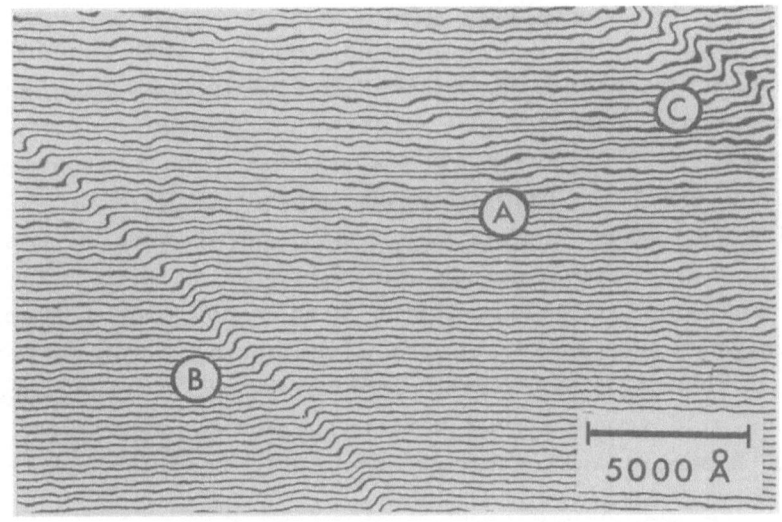

Fig. 5.6 Interference micrograph of Molybdenite
thin film (phase amplification: x24).

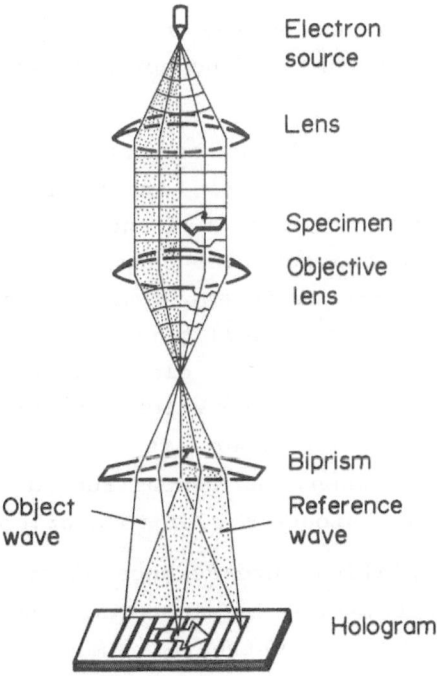

Fig. 5.7 Schematic diagram for electron-hologram formation.

5.1.2 Experimental Method

In the experiment by Tonomura et al. [139], [184], electron holograms were formed in a 100kV field-emission electron beam so coherent as to produce 3,000 biprism interference fringes (this is in contrast to 300, which is the highest number attainable with a thermionic electron beam). The schematic diagram for hologram formation is shown in Fig. 5.7. A toroidal sample is situated in one half the specimen plane; the other half is for the reference beam. The sample is illuminated with a collimated electron beam. The divergence angle, 2β, of the illuminating electron beam must be so small that the transverse coherence length (the length within which the wave front can be defined), given by $\frac{\lambda}{2\beta}$, is large enough to cover both sample and reference beam. In this experiment, $2\beta < 10^{-7}$ rad., and $\frac{\lambda}{2\beta} > 0.03(\overset{\circ}{A})/10^{-7} = 30$ (μm).

The object beam (phase-shifted by the sample) and the reference beam are brought together by the electron biprism located below the objective lens to form an interference pattern. The pattern is enlarged 1,000 times by electron lenses and recorded on film as a hologram.

The phase shift caused by the sample was reconstructed with a He-Ne laser light (λ': 6328$\overset{\circ}{A}$) in the optical system shown in Fig. 5.8.

There, the collimated light beam from the laser is incident into a Mach-Zehnder interferometer, which is composed of two half mirrors and two mirrors located at diagonal apexes of a rectangle. The beam is split into two coherent beams by the first half-mirror. The beams travel along two different arms (A and B) of the rectangle to be recombined by the second half mirror. The two beams, which are almost parallel, illuminate the electron hologram. Each beam produces two diffracted beams, one which reconstructs the phase shift due to the sample, and the other its conjugate. An interference micrograph is obtained when the reconstructed image of beam A and beam B are made to pass through a slit to overlap with each other.

A twice phase-amplified interference micrograph can be obtained using the same optical system by adjusting the Mach-Zehnder inter-ferometer so that the reconstructed image of beam A and the conjugate image of beam B, or vice versa, are superimposed. The amplitude of the conjugate image, which inevitably appears in a holographic reconstruction process, is a complex conjugate of that for the reconstructed image. Therefore, the phase difference between the two images is two times as large as that between the reconstructed image and a plane wave.

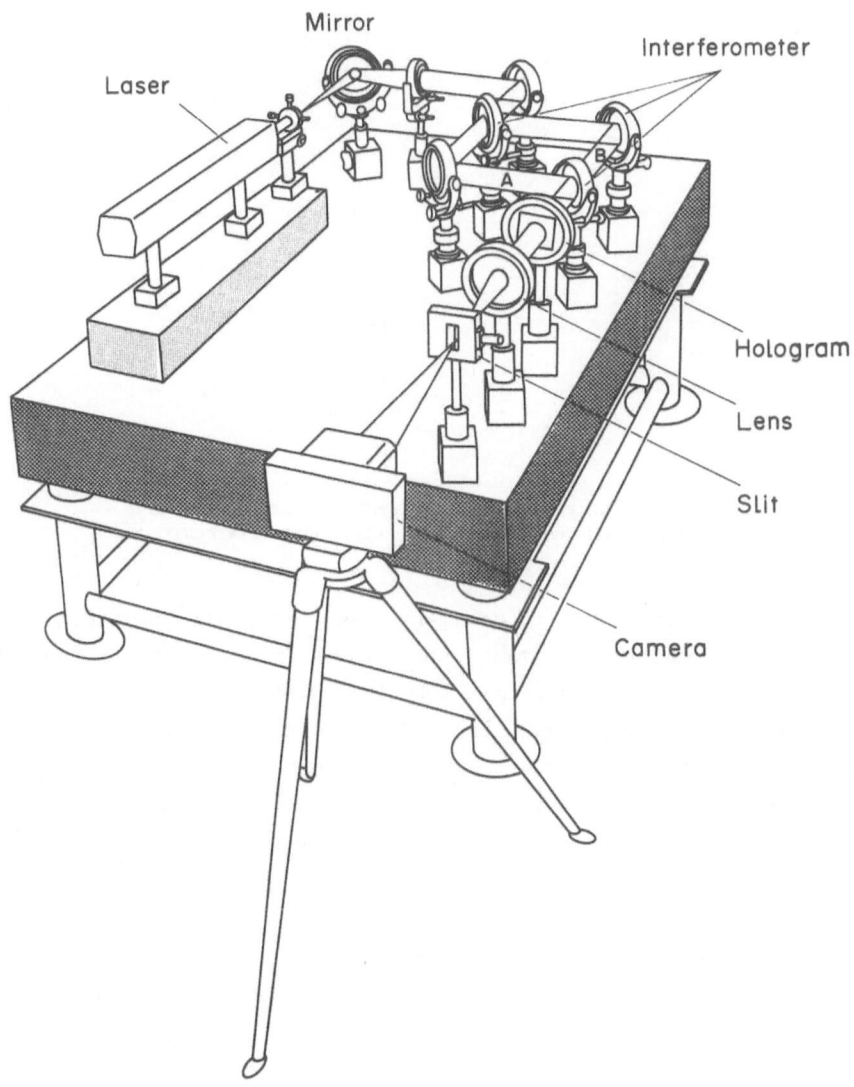

Fig. 5.8   Optical reconstruction system for
            interference microscopy.

Let me mention here that interference micrographs can generally be classified into two categories: contour maps and interferograms. The former type is obtained when the directions of the reconstructed object wave and a reference plane wave are the same in the optical reconstruction system (see Fig. 5.9(a)). In such maps, contour lines of the wave front can be seen.

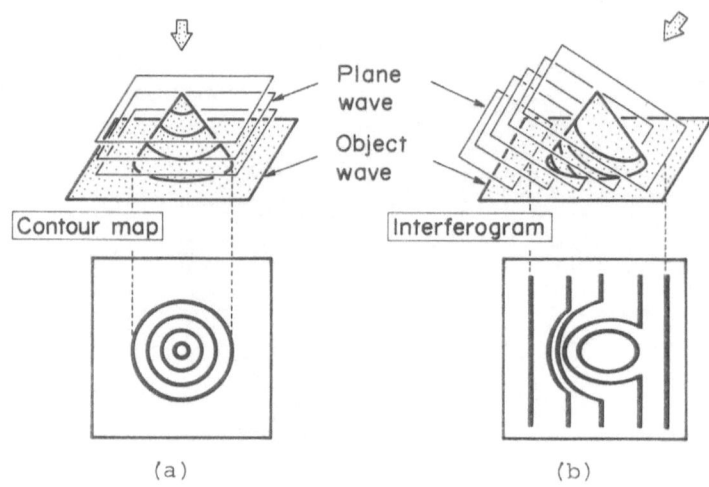

Fig. 5.9   Two kinds of interference micrographs:
(a) Contour map, and (b) interferogram.

When only using such a contour map, it unfortunately is not possible to determine whether the wave front is protruded or retarded. Such protrusion-related information can be obtained, though, using an interferogram for when the direction of a reference plane wave is not the same as that of the object wave (Fig. 5.9(b)). An appropriate analogy here might be that a mountain can be recognized more easily when viewed obliquely rather than from directly above.

An example can be seen in Fig. 5.5(c) and (d). Such interferograms were actually used in this experiment to measure relative phase shifts between two beams passing inside the hole and outside the toroidal sample.

5.1.3 Sample Preparation

In the 1982 Tonomura et al. experiment, tiny ferromagnetic samples of square toroidal geometry were fabricated photo-lithographically in the following way. Thin films of Permalloy (80%Ni and 20%Fe) 400Å thick were prepared through vacuum evaporation. The substrate was configured of a glass plate covered with an evaporated thin film of NaCl. Toroidal samples were then cut out of the Permalloy film as described in the next paragraph. (see Fig. 5.10).

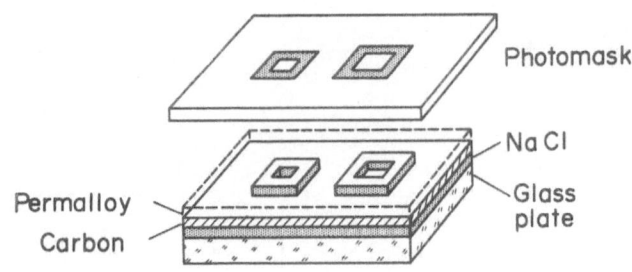

Fig. 5.10   Schematic for fabricating square
             toroidal samples.

First, a photoresist was coated over the Permalloy surface, and the toroidal photomask pattern was transferred to the resist. After the development process, the only resist remaining on the Permalloy film was in a toroidal shape. The Permalloy film was then ion-milled, except for the area covered with the resist. This photoresist left on the Permalloy toroids was removed by fixing. The toroids were floated off on a water surface and applied to thin carbon film approximately 100Å thick. The magnetic flux flowing inside such toroids ranged from 3(h/e) to 10(h/e), as estimated from the widths of the toroids. Leakage fields from the toroidal samples were measured by holographic interference electron microscopy. They were confirmed to be too small to have any influence on the experimental conclusion.

Two typical interference micrographs of the toroidal samples with and without any leakage flux are shown in Fig. 5.11.

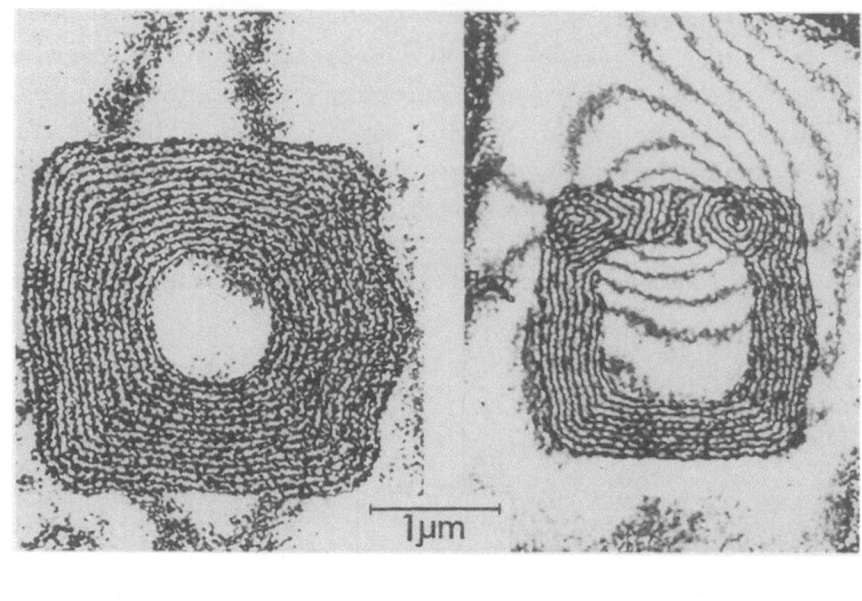

(a)   (b)

Fig.5.11   Interference micrographs of toroidal
           ferromagnets:(a) Closed magnetic circuit,
           and (b) open magnetic circuit.

Since the micrographs are phase-amplified two times, a constant
magnetic flux of h/2e flows between two adjacent contour fringes.
In micrograph (a), almost all of the flux flows inside the magnet.
More precisely, it can be seen that only 1/15 of the total flux,
i.e. h/2e, leaks outside.   This means the leakage can affect the
phase of an incident electron beam by as little as $2\pi/2$.   In micrograph (b), however, all of the flux leaks outside the toroid, perhaps
due to some imperfections in the toroidal shape.   Only samples found
to be without any leakage flux were selected for the experiment.

5.1.4 Experimental Results
     The phase distribution of an electron beam transmitted through
a toroidal sample was observed as an interferogram.   It was
concluded that if there was no relative phase shift between two

electron beams passing both through the hole in the toroid and outside it, the interference fringes must be along a common straight line inside and out. The sample interferogram shown in Fig. 5.12 clearly indicates the existence of a relative phase shift of approximately 12 π.

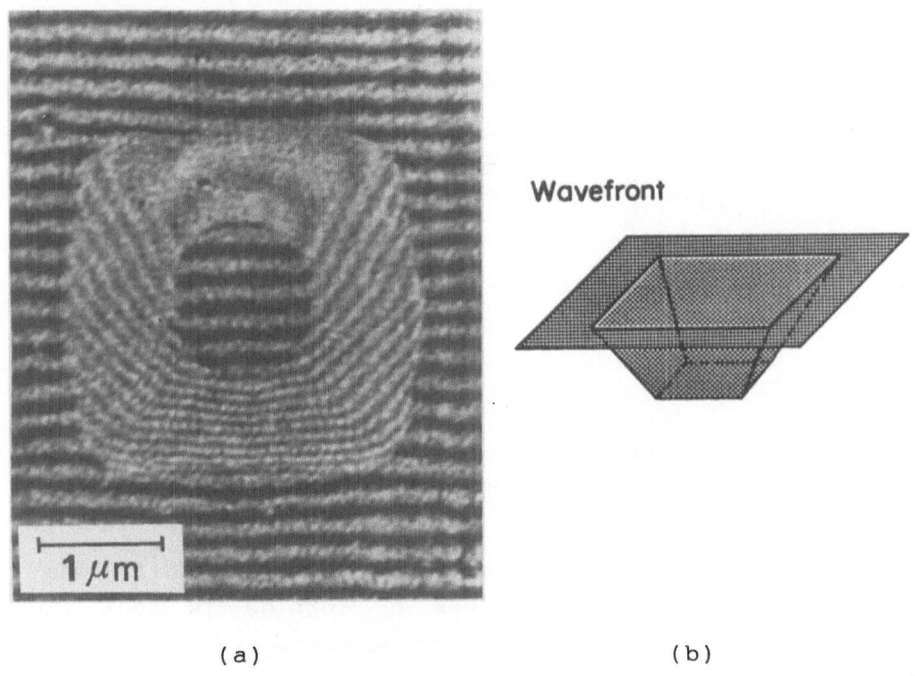

Wavefront

(a)   (b)

Fig. 5.12   Interferogram of square toroidal ferromagnet:
(a) Interferogram, and (b) calculated wave
front image.

That is, the transmitted wave front displays a relative displacement between the beams passing inside and outside the toroid.

It is possible that such a phase shift can also be produced by a change in specimen thickness. One therefore might ask if the observed phase shift could be due to any change in thickness. This possibility is negated, however, when we look at Fig. 5.13 to see an interferogram of another sample. There, the wave front displacement is reversed. This is due to the AB effect when the magnetization direction in the toroid is reversed; it cannot be explained by the thickness change.

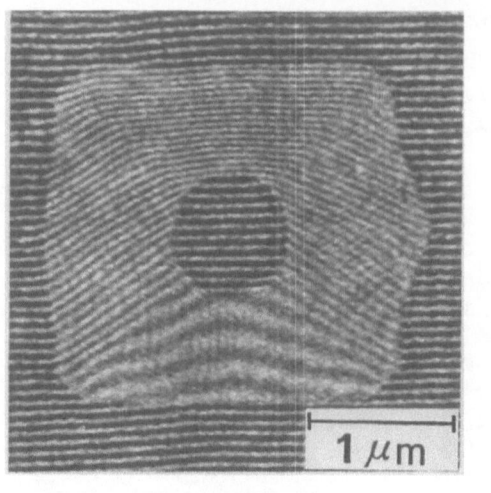

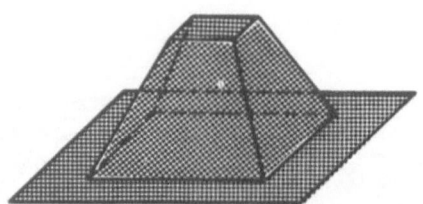

(a)   (b)

Fig. 5.13   Interferogram of another toroidal ferromagnet:
             (a) Interferogram, and (b) calculated wave
             front image.

Quantitative confirmation of the AB phase shift was also carried out. Since the thickness of toroids is 400Å and the magnetic induction (4 π times the magnetization) is equal to 9500 Oe, the magnetic flux flowing through the toroid in Fig. 5.12 (average width of 6400Å) is 2.4 x 10^{-14} Wb or 6.0 (h/e). This is consistent with the observed 12 π phase shift. The toroid in Fig. 5.13 (toroid width of 8600Å) has a flux of 8.0 (h/e), which can also be confirmed in the fringe shift.

This quantitative agreement is characteristic of the AB effect. That is, the phase shift value is completely determined by the enclosed magnetic flux, and does not depend on any other experimental condition such as electron energy. It should be noted in passing that a phase shift due to a thickness change depends on the electron energy. All these results support the existence of the AB effect.

In the experimental arrangement discussed here, the incident electron beam touched and even penetrated the magnet. This was because penetrable toroidal magnets were adopted for measurement of the absolute phase shift. If the fringes inside the image of toroid had not been observed, the phase shift would have appeared to have been determined modulo 2π.

The validity of the experiment is proven by the fact that the part of the beam transmitted through the magnetic flux in the sample does not contribute to the beam traveling outside the sample. In fact, the shape of the sample is formed through an electron lens as an in-focus image on the interferogram. This was possible because all of the electrons starting from an object point were focused through an electron lens into an image point. In other words, a phase distribution at the specimen plane was reproduced at the image plane of the electron lens.

5.1.5 Discussions on Validity of Experiment

The validity of this experimental test [139] of the AB effect was questioned by Bocchieri, Loinger and Siragusa [188], as well as by Miyazawa [189].

Bocchieri, Loinger and Siragusa [188] asserted that the effect observed by Tonomura et al. [139] could be interpreted by taking into account the action of the Lorentz force on the portion of the electron wave which penetrated into the magnetic field. They pointed out that penetration of a portion of an electron wave into a magnetized region generally induces a sudden adjustment of the whole electron wave. This adjustment, they said, precisely simulates the AB effect. Therefore, they concluded that this experiment provides no proof of the AB effect. They remarked upon the necessity of designing an experiment for testing the AB effect where the electron waves would traverse regions completely free of magnetic fields.

Miyazawa [189] also argued against Tonomura et al. This experiment, he said, did not effectively test for the existence of the AB effect. In order to do so, he stated that the wave function of an incident electron beam must be zero at all the boundaries of the magnet. If such boundary conditions were to be satisfied in a new experiment, he was confident the so-called AB effect would vanish. His reasoning was as follows.

He first discussed whether or not the wave function is single-valued. Multi-valued wave functions, he said, cannot be excluded a priori, as had been done by several authors (For example, Schrödinger [190], Pauli [191], Blatt and Weisskopf [192]). The single-valuedness of the wave function in connection with the AB effect was also investigated by Merzbacher [193], Tassie and Peshkin [26], Kretzschmar [194], and Silverman [195]). Miyazawa adopted the minimum energy principle after giving up on the postulate of single-valuedness regarding the AB effect for bound-state electrons orbiting a magnetic string. He concluded that the wave function is single-valued, producing the so-called AB effect when the electron wave function and the magnetic field overlap. However, his interpretation was that the wave function becomes multi-valued, causing the dis-appearance of any AB effect when the overlap decreases. That is, the electron makes a transition from one state to the other when this overlap changes.

Comay [196] pointed to the possibility that a toroidal single-domain ferromagnet employed in the experiment by Tonomura et al. [139] might have produced an additional electrostatic potential around the magnet due to the electric quadrupole moment of the partially occupied electronic d-shell. This potential would produce a phase shift for an incident electron beam in the same way that the AB effect might. Although the theoretical value of this phase shift was smaller than the observed one by two orders of magnitude, he asserted that the effect of the potential could be detected by clarifying the difference between two measurements in which the toroid was put face up and face down. Actually, this method had already been utilized to distinguish between electric and magnetic effects by Tonomura et al. [197] in electron holography aimed at determining the three-dimensional magnetic domain structure of a fine particle.

Home and Sengupta [78] emphasized the necessity of using a toroidal magnet when testing for the AB effect. They asserted that the electron penetration effect in Tonomura et al.'s experiment should be probed more definitively.

In 1986, Walker [198] carried out an experiment concerning the AB effect, where he utilized superconductors instead of electron beams. He pointed out that the conclusions drawn from the experiment by Tonomura et al. [139] were open to debate since some of the electrons in the beam actually passed through the magnet. Walker's experimental arrangement was essentially the same as that by Jaklevic et al. [13], except for the use of a toroidal solenoid instead of a straight one. The AB effect was detected by measuring a periodic SQUID current with a flux period of h/2e.

Experimental conditions precluded objections arising from the use of
linear flux geometries, or the use of an experimental arrangement
which would allow electrons to penetrate the region containing
magnetic flux.

5.2 Experiment Using Solenoid Covered with Glass Tube [199]

In 1982, Möllenstedt, Schmid and Lichte [199] carried out an
experiment to test Bocchieri and Loinger's assertion [77] that even
a weak electron wave penetrating into the solenoid could produce the
phase shift. Bocchieri and Loinger asserted that this effect would
explain previous observations by Möllenstedt and Bayh [11]. In the
1982 experiment, a solenoid was enclosed in a glass tube (diameter:
80 μm) covered with a gold layer. An incident electron beam was
split by biprism 1 into two beams with a separation of 200 μm. (see
Fig. 5.14).

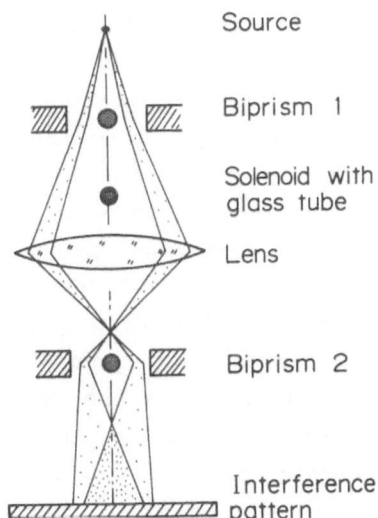

Fig. 5.14   Experiment by Möllenstedt et al. using
            solenoid covered with glass tube.

The two beams enclosed the solenoid, and were brought together with
biprism 2 to form an interference pattern.

The phase distribution of the electron beam that passed beside
the glass tube was seen to remain unchanged when the current through

the solenoid was turned on. In other words, no magnetic field was produced near the solenoid. The relative phase shift between two electron beams enclosing the solenoid was then measured as coming within 3% of the theoretical value.

5.3 Experiment to Test Possible Flux Quantization [200]

If, by some remote chance, the magnetic flux in an "autistic" magnet could be quantized as predicted by Costa de Beauregard and Vigoureux [163], certain researchers in the field felt that this would have a great influence on interpretations of the AB effect. An impenetrable toroidal magnet containing a magnetic flux quantized in h/e flux units would produce no physical effects on an electron beam incident around the magnet.

Tonomura et al. [200] experimentally tested this possibility. The experimental method was essentially the same as that described in Section 5.1, except for higher precision in the flux measurement. In the previous experiment using square toroidal magnets, the amount of leakage flux was approximately h/2e -- thus the possibility of flux quantization in h/2e units could not be discussed.

In order to decrease leakage flux considerably below h/4e, the shape of toroidal magnets was made circular instead of square. In addition, the total magnetic flux flowing through each toroid was made as small as h/e. Such toroids were fabricated by a process differing slightly from the previous one. A glass substrate that was previously covered with double layers of evaporated NaCl, and carbon was coated with photoresist (Fig. 5.15). Negative patterns were formed in the photoresist, and then Permalloy films $100-500\overset{\circ}{A}$ thick were evaporated on them. When the photoresist was dissolved, only the positive patterns of Permalloy remained on the substrate. These patterns on the carbon film were floated off on a warm water surface by dissolving the NaCl layer. They were then placed on a supporting mesh.

These samples were first examined by holographic interference microscopy, and only samples with a leakage flux of less than h/10e were selected. The possibility of flux quantization was checked by observing two-times phase-amplified interferograms of toroidal samples.

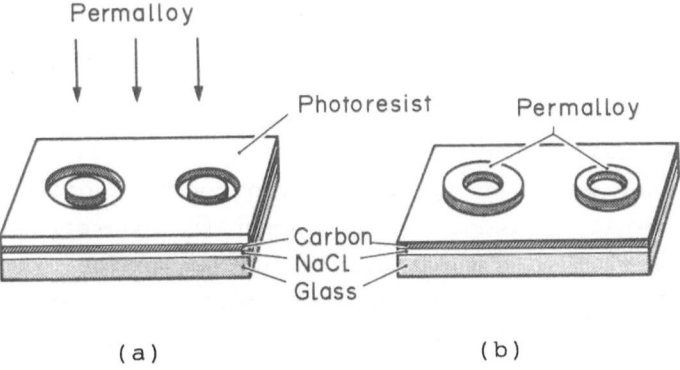

(a)   (b)

Fig. 5.15 Schematic for fabricating circular
           toroidal samples: (a) Negative
           photoresist pattern, and (b)
           positive Permalloy pattern.

It was concluded that magnetic flux in a toroidal magnet was
quantized neither in h/e units nor in h/2e units. A typical example
of an interferogram indicating non-quantized flux is shown in Fig.
5.16.

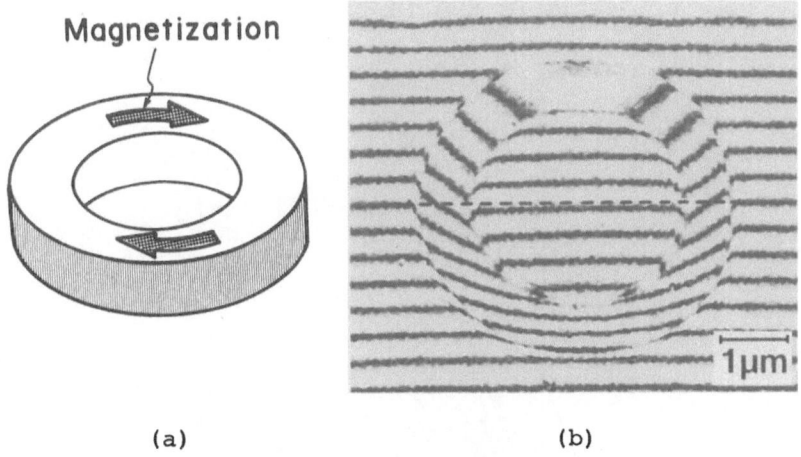

(a)   (b)

Fig. 5.16 Interference micrograph of circular toroidal
           ferromagnet:(a) Schematic, and (b) interferogram.

This conclusion gives further evidence for the existence of the AB
effect.

5.4 Experiments Concerning Electrostatic AB Effect [201]

Although the magnetic AB effect has often been experimentally investigated, the electric AB effect (see Fig. 2.1) has never been successfully clarified. This is due to the technical difficulties involved in experiments on the electric AB effect. Here, an electron wave packet would be split into two coherent packets, which would enter two metal cylinders. Only when the packets were well inside the cylinders would the potentials applied to the cylinders be changed. Although the electron would be subject to no force, a phase shift would be produced. This is what is meant by the electrostatic AB effect.

A phase shift related to the classical time lag of an electron beam was investigated by Matteucci and Pozzi [201]. In this experiment, an electron beam is incident around a filament, the surface of which is covered with two different metals as shown in Fig. 5.17.

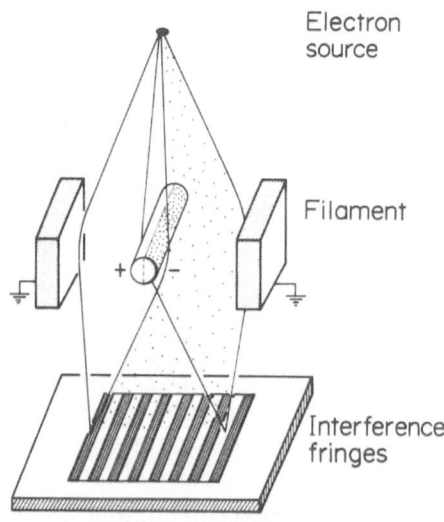

Fig. 5.17   Experiment by Matteucci and Pozzi concerning electrostatic AB effect.

Due to the difference in contact potential between the two metals, the two beams on opposite sides of the filament pass through regions having different potentials, forming a diffraction pattern in the lower plane. Although the electron beams experience electrostatic forces near the filament, this produces no net change of energy but only a classical time lag. The relative phase shift was confirmed by observing a change in the diffraction fringes when the filament was rotated. Such a phase shift had been discussed by Boyer [45] and Aharonov [177] in relation to the AB effect.

Another piece of pertinent work is Schmid's experiment [202] aimed at determining the coherence length of an electron wave packet. A 35kV electron beam was split into two beams. Only one beam was made to pass through a metallic tube having electrostatic potential U. Finally, the two beams were recombined to form an interference pattern (see Fig. 5.18).

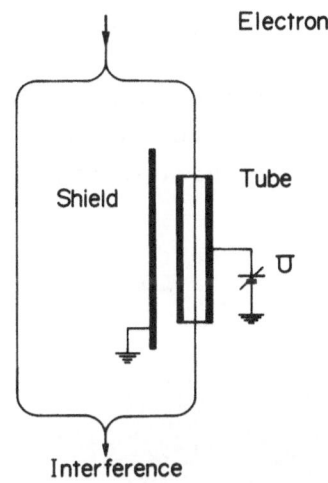

Fig. 5.18 Schmid experiment to determine coherence length of electron wave packet.

By increasing U, the maximum observable phase shift was measured. This maximum phase shift was found to be 70,000 x 2π, which corresponds to a coherence length of 4,600Å.

5.5 Experiment Using Toroidal Magnet Covered with Superconductor [203-205]

Tonomura et al. [139] asserted that the existence of the AB

effect was experimentally confirmed by using a magnet of toroidal geometry which had no returning fields due to the completed magnetic circuit. In addition, Möllenstedt et al. [199] confirmed that the electron penetration effect into the solenoid was too small to overthrow the previous conclusion arrived at using a solenoid covered with a glass tube.

However, these results still did not convince opponents of the AB effect. The opponents asserted, on the one hand, that the detected phase shift could be interpreted as a result of the Lorentz force on electrons which partly penetrated the magnet. On the other hand, they said, the leakage flux must be reduced to zero for a true test.

In 1986, Tonomura et al. [203] again carried out an experiment, which fully satisfied the required conditions for an experimental test of the AB effect. A description of the experiment follows.

5.5.1 Design of Experiment

First, a toroidal magnet had to be completely shielded with a metal layer from an incident electron beam so as to preclude the electron penetration effect as shown in Fig. 5.19.

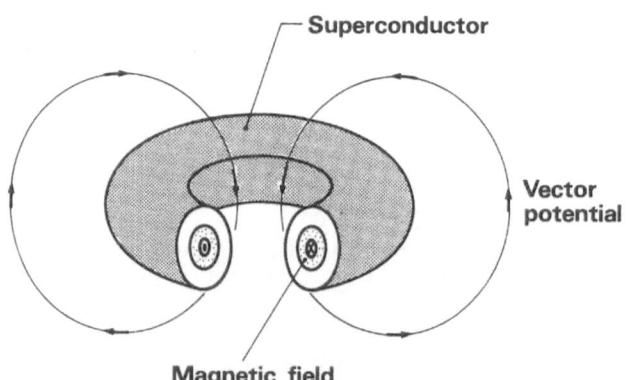

Fig. 5.19   Schematic of toroidal ferromagnet covered
with superconductor.

Although one might think that a shield layer only on the top surface of a magnet would serve the purpose, such an experiment had already been carried out, and the AB effect had been detected. (see reference [204]). Moreover, it had been asserted by Miyazawa [189] that the entire surface of a magnet should be covered in order to satisfy the boundary condition that the wave function vanishes at the surface of the magnet.

By following this approach, the sample becomes free from electron penetration. The price to pay, however, is that the absolute phase difference cannot be measured except modulo 2π. Another problem is that although an electron beam cannot penetrate the magnet, a small amount of leakage flux from a magnet might possibly influence the electron beam.

In order to prevent this possibility, Tonomura et al. made a shielding metal layer of superconducting material. Magnetic fields could not pass through the superconducting layer due to the Meissner effect. Consequently, no magnetic fields leaked outside the toroidal sample. The fabricated sample could be proven to have no leakage flux utilizing electron holography. Accordingly, the structure of the sample was determined as shown in Fig. 5.19.

It thus became possible to test for the AB effect using this sample under conditions where there was no overlap between the magnetic field and electron beam. Furthermore, this scheme also could serve as an idealized test of Liebowitz's assertion [34] that the AB effect can be interpreted in terms of force concepts. This was possible because the electron's magnetic field could not penetrate into the region of the magnetic field due to the covering superconductor. Such a test was originally proposed by Erlichson [40], and was carried out experimentally using a finite superconducting cylinder trapping a magentic flux by Lischke [42].

5.5.2 Sample Preparation

The first problem to overcome in sample preparation was a technical one: how could such a complicated sample be fabricated? A tiny toroidal magnet of less than 10 μm diameter had to be completely covered by a superconductor, without even the slightest gap. The thickness of the superconducting layer should be greater than the penetration depth. Samples satisfying such conditions were actually fabricated using advanced photolithographic technique. The fabrication process was a bit complicated, thus only its principle is shown in Fig. 5.20.

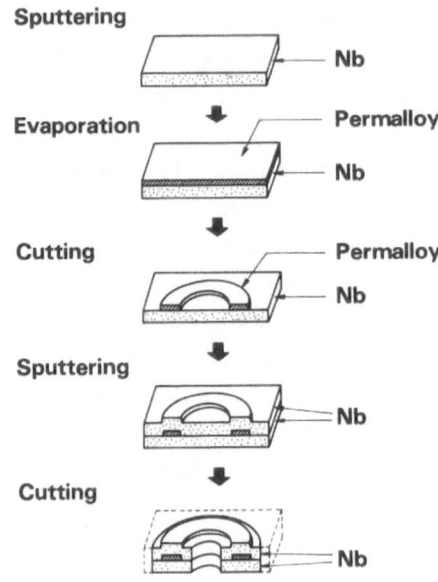

Fig. 5.20   Processes involved in fabricating toroidal
ferromagnet covered with superconductor.

A Permalloy film, 200Å thick, is prepared by vacuum evaporation
on a silicon wafer covered with a niobium film 2500Å thick. After the
toroidal shape was cut out of the Permalloy film, a 3000Å niobium
film was sputtered on its surface.   Then, the toroidal sample was
cut so that the Permalloy toroid was completely covered with the
niobium layer.   Finally, a copper film 500 – 2000Å thick was
evaporated on all its surfaces; the film prevented penetration of
any electron beam into the magnet and kept the sample from
experiencing charge-up and contact-potential effects.   A scanning
electron micrograph of the sample is shown in Fig. 5.21.

Special attention was paid to attaining perfect contact between
the two niobium layers; even a slight oxidation layer between them
would break the superconducting contact and allow magnetic field
leakage outside.

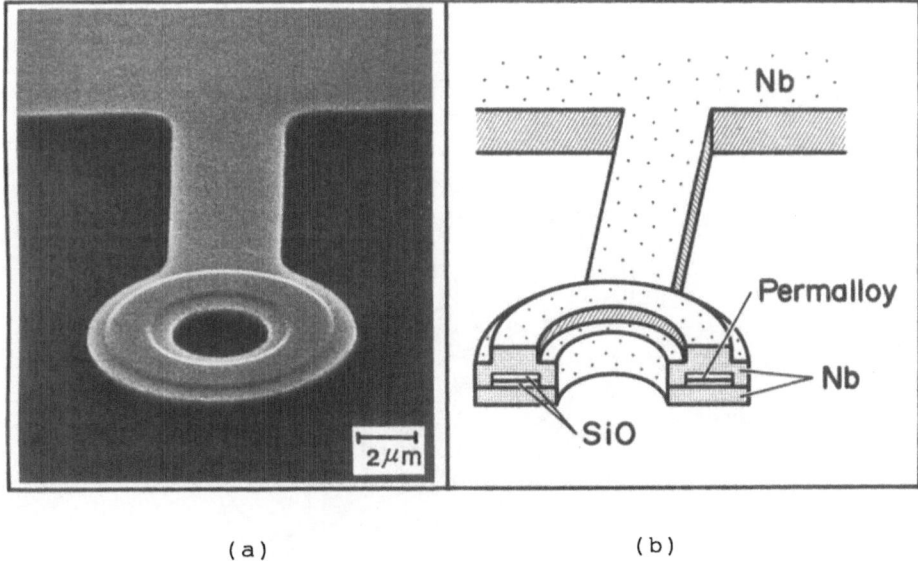

(a)   (b)

Fig. 5.21   Toroidal ferromagnet covered with niobium layer:
            (a) Scanning electron micrograph, and (b) cross-
            section.

Therefore, the niobium oxide produced at the top surface of the
lower niobium layer in the lithography processes had to be removed
by ion sputtering before laying down the upper niobium layer.   A
maximum current density of 40 mA/μm^2 was confirmed in another
experiment to pass through the contact at 4.2K.   This current was
much larger than the persistent current necessary for quantizing the
magnetic flux ($\sim$10 mA/μm^2).

5.5.3 Low-temperature Stage
     The second problem was also a technical one.   How could the
sample be cooled well below the 9.2K critical temperature of
niobium?   A special specimen stage was developed for this experiment; its cross-section is shown in Fig. 5.22.   Each specimen was
completely enclosed by three-fold radiation shields.   They were
connected through conduction rods to three reservoirs.   The outer
one was filled with liquid N_2, and the inner two with liquid He.

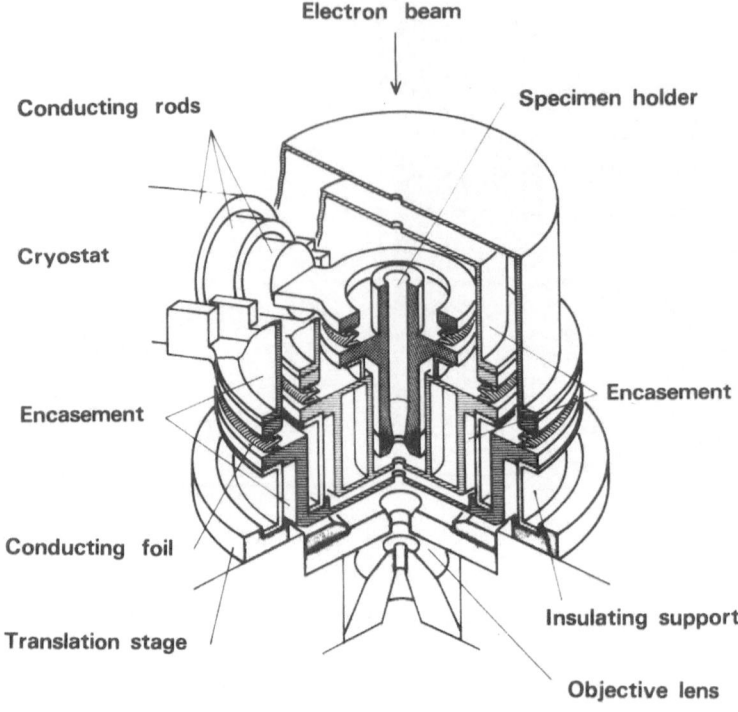

Fig. 5.22 Low-temperature specimen stage.

The temperature of the specimen was controlled by heater over a range between 2K and 50K. Measurement was carried out using a Ge resistor. Directing an electron beam onto a sample raised its temperature. However, it was possible to limit any rise to less than 0.5K with normal illumination. This was done by connecting the sample to a large niobium plate via a tiny bridge, as can be seen in Fig. 5.23. In this way, high thermal conductivity was achieved between the sample and the liquid He reservoir.

Charging-up of a sample as a result of electron irradiation often occurred during cooling of the sample. This produced an unwanted electrostatic potential distribution, which contributed to a major drop in phase measurement precision. More specifically, the charging-up was due to electron irradiation of an insulator layer deposited on the sample during the cooling process.

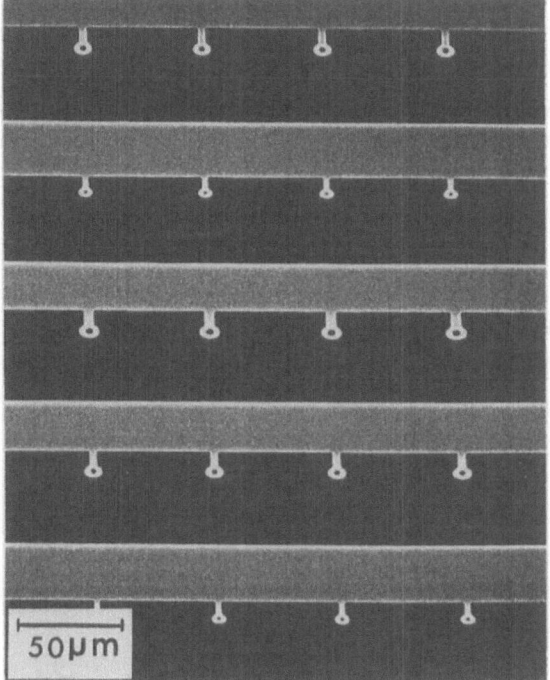

Fig. 5.23 Toroidal ferromagnets.

To prevent a sample from charging-up because of such a deposit, careful attention was paid to the cooling procedure. The sample temperature was kept as high as possible until the last moment of measurement.

5.5.4 Experimental Method

Experiments were carried out in a manner similar to that described in Section 5.1. First, an electron hologram of a sample was formed with a 150kV field-emission electron microscope. Then, the relative phase shift for the sample was optically reconstructed using a He-Ne laser. At the same time, the leakage magnetic flux from the sample was quantitatively measured from the optically obtained interference micrograph.

5.5.5 Interaction of Electron with Superconductor

Before measuring the relative phase shift, a preliminary experiment was carried out to test whether or not there would be any observable interaction of an electron beam with a toroidal superconductor containing no magnet. The reasoning behind this test was as follows. An incident electron passing through a field-free region has no reason to receive any force. However, when a superconductor is located near the electron, the electron might somehow be influenced because the magnetic field produced by the passing electron cannot penetrate into the superconductor, thanks to the Meissner effect.

Samples used in this experiment were niobium toroids, 3000Å thick, their surfaces completely covered by an evaporated copper layer 500 ⌄2000Å thick. The phase difference was measured between two electron beams passing both inside and outside the toroid while the temperature of the toroid was varied above and below the critical temperature of $T_c (=9.2K)$.

Examples of interferograms at 15K and 4.5K are shown in Fig. 5.24.

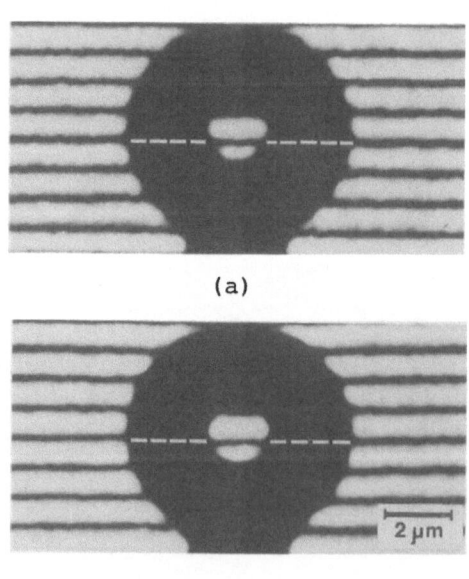

(a)

(b)

Fig. 5.24   Interferograms of superconducting
toroid: (a) T=15K, and (b) T=4.5K.

No phase difference could be observed in the twice phase-amplified interference micrograph for both normal and superconducting cases.

This experiment confirmed that there was no interaction between the electron beam and superconducting toroid within the precision of a $2\pi/10$ phase shift. In order to acquire experimental evidence that the niobium layers actually became superconductive in this experiment, another experiment had to be performed. In it, magnetic flux was applied to the sample above T_c, and the trapped flux in the sample below T_c was detected by electron holography.

5.5.6 Experimental Test of the AB Effect

The preliminary experiment confirmed that no relative phase shift was produced between two electron beams passing both inside and outside a superconducting toroid containing no magnet. As the next step, the AB effect was then tested using a toroidal magnet covered with a superconductor. A sample was fabricated in the manner described in Subsection 5.5.2. The fact that the magnetic circuit was closed was confirmed by measuring leakage flux of the sample using holographic interference microscopy.

The fabricated samples were not always free of leakage flux. On the contrary, leakage was found to be appreciable for sample geometries with large diameters (e.g. larger than 10 µm), or small aspect ratios. After the sample geometry was redesigned, approximately 60% of the samples were confirmed to be free from leakage within a precision of $(1/20)(h/e)$. An example of a sample with a large leakage flux is shown in Fig. 5.25.

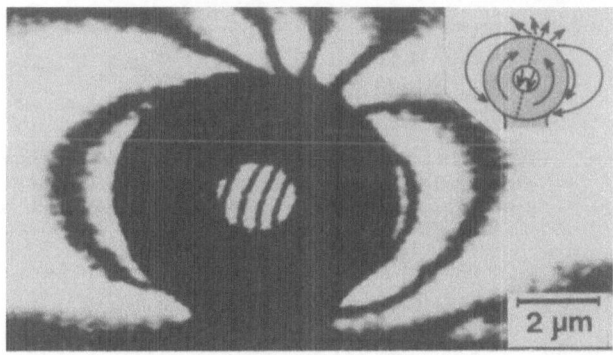

Fig. 5.25   Leakage flux from toroidal ferromagnet
(phase amplification: x2).

A leakage-free toroid was selected as a sample and cooled down to 5K on a low-temperature stage. Then, an electron beam was shined onto the sample to measure the relative phase shift between two partial beams passing inside and outside the toroid. Although samples having various values of magnetic fluxes were measured, only two kinds of interferograms were obtained (see Fig. 5.26(a) and (b)).

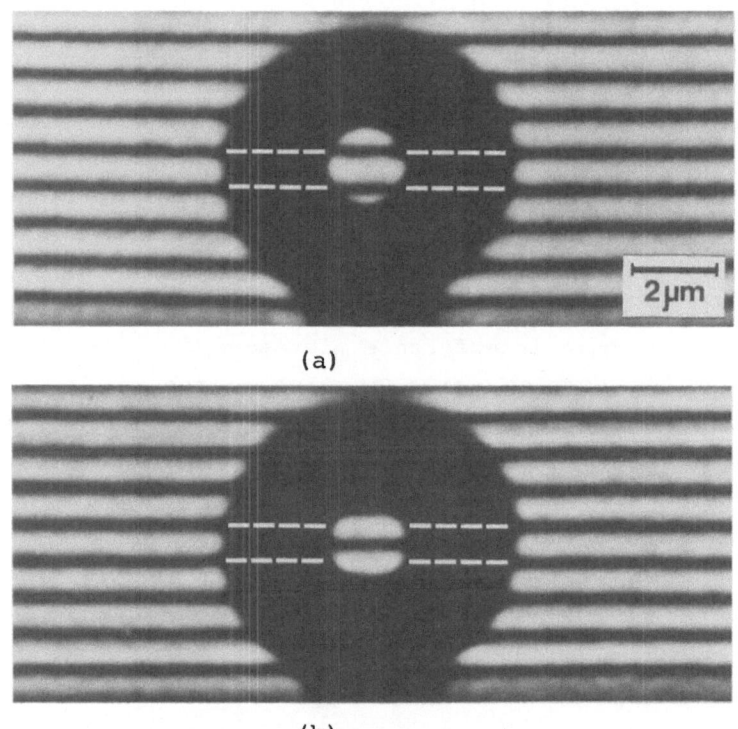

(a)

(b)

Fig. 5.26   Interferograms of toroidal ferromagnets covered with superconducting layers at 4.5K: (a) Phase shift = 0, and (b) phase shift = π.

That is, the value of the phase shift ended up being either 0 or π. Photograph (b) makes it clear that a relative phase shift is clearly produced between two beams passing through field-free regions. This proves that the AB effect exists. Let me emphasize here that the magnetic field was confined within the superconductor, and that the field was shielded from the electron beam by the copper and niobium covering.

The existence of a π phase shift is clear from photograph (b). Why, however, did the phase shift become either 0 or π? In order to answer this question, further experiments were required.

5.5.7 Temperature Dependence of Phase Shift

A temperature dependence experiment for the phase shift was used to find out why the measured phase shift had a discrete value of only 0 or π. This quantization of the phase shift provided key evidence for the complete shielding of a magnetic field with the covering superconductor; thus, it would be good to look at it in greater detail.

The mechanism for phase shift quantization was clarified by the following interference experiment. When the temperature, T, of the toroidal sample was raised, the interferogram changed abruptly when T exceeded the superconducting critical temperature, T_c. The phase shift value remained either 0 or π , independent of T, as long as the magnetic flux was covered with a superconductor, i.e. $T < T_c$. However, when $T > T_c$, the phase shift value could be considered to be determined only by the magnetic flux of the toroidal Permalloy. This was proven by the experimental fact that the phase shift value decreased in proportion to a decrease in magnetization of the Permalloy when the sample temperature was further raised to room temperature.

An example is shown in Fig. 5.27. Here, the sample is the same as that shown in Fig. 5.26(b). Relative displacement of the fringes inside and outside the toroid is just half a fringe spacing at 4.5K (see Fig. 5.27(a)). When the temperature is raised to 15K, thus exceeding the critical temperature of 9.2K, fringe displacement shifts to an 0.4 fringe spacing, or an 0.8π phase shift, as can be seen from Fig. 5.27(b). Here, the sample is the same as that shown in Fig. 5.26(b).

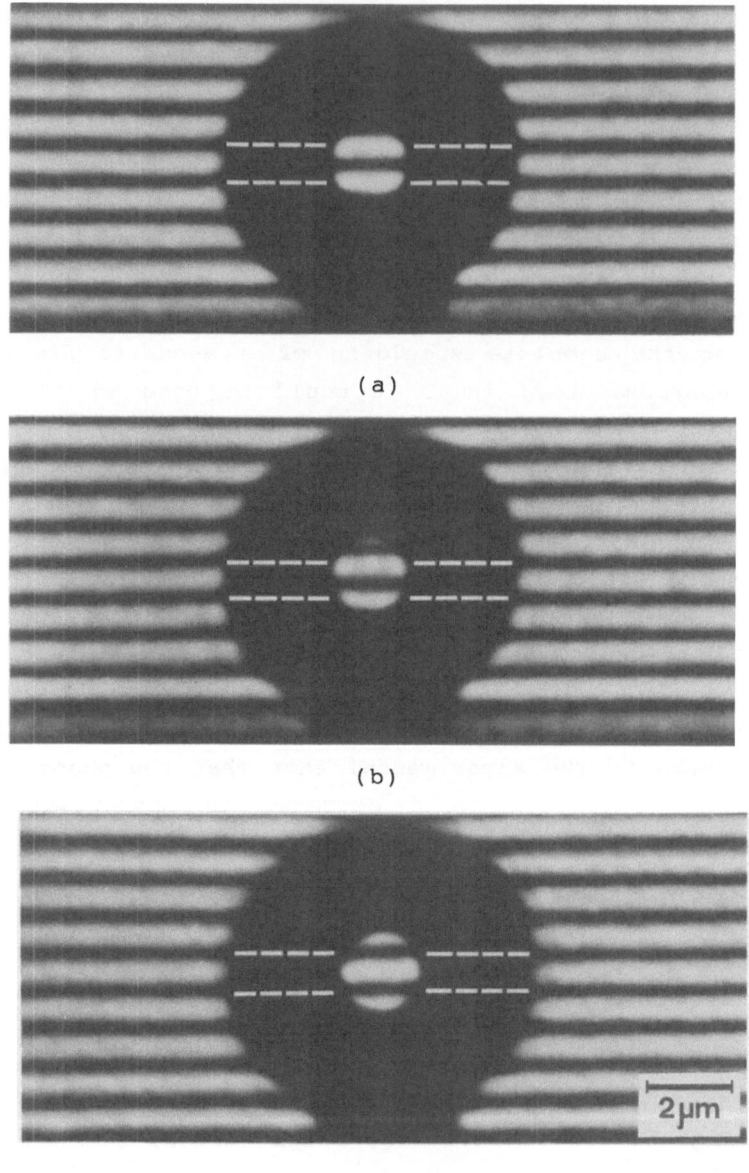

Fig. 5.27 Temperature dependence of interferogram
of toroidal ferromagnet: (a) T=4.5K,
(b) T=15K, and (c) T=300K.

It is now clear why the quantized phase shift value in this sample was π rather than 0. A phase shift of π was selected simply because the phase shift just above T_c was 0.8 π, which was accidentally closer to π.

When the temperature of the sample was further raised to room temperature, the fringe displacement decreased to an 0.15 fringe spacing as shown in Fig. 5.27(c). The reason for this change has already been explained. The magnetization and therefore the magnetic flux decrease due to the increase in thermal fluctuation of the spins together with a rise in temperature. Since the magnetization of the Permalloy decreases by 5% when the temperature changes from 15K to 300K (see Osakabe et al. [205]), the decrease in magnetic flux (∿ 5(h/e)) is 0.25 (h/e), which clearly explains the change in fringe displacement from an 0.4 to 0.15 spacing.

It is possible to conclude that the phase shift above T_c is completely determined by the magnetic flux of the Permalloy, and that the phase shift below T_c shows a quantized value which shifts between the two values of 0 and π according to a selection rule. This rule can be expressed as "whichever of the two values is closer to the phase shift value just above T_c will be chosen".

5.5.8 Magnetic Flux Quantization

We have finally arrived at the essential problem of what happens to a toroidal sample below T_c, and why the phase shift must have a discrete value of 0 or π. The question can be answered based on the experimental results described above.

When there are no magnetic fields the phase of the Cooper pair wave is constant throughout the superconductor. However, when the superconductor encloses a magnetic flux, as in the present case, the phase of the wave becomes a function of a position. This is because the phase shift of a Cooper pair can, quite similarly to the case for electrons, be given by

$$\Delta S = \frac{1}{h} \oint (2M\mathbf{v} - 2e\mathbf{A}) \, d\mathbf{s}. \quad \ldots\ldots\ldots\ldots\ldots\ldots\ldots (5.1)$$

It follows from this equation that even if there are no magnetic fields inside the superconductor, the phase can be shifted by the existence of vector potentials. This may be called the AB effect for Cooper pairs. The AB effect is, at least in principle, limited not only to electrons but also to any kind of charged particles.

When the wave turns once around the magnetic flux and returns to the original position (see Fig. 5.28), the phase has to be the same as the original value, modulo 2 π.

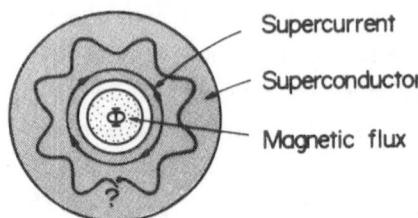

Fig. 5.28 Principle underlying magnetic flux quantization.

Otherwise, the Cooper pair wave will vanish, and the superconducting state will consequently revert to the normal state. This means the phase shift, ΔS, integrated along a closed path around the magnetic flux should be a precise integral multiple of 2 π, i.e. 2n π. However, the phase shift - $2e \oint A \, ds/\hbar$ is not necessarily equal to 2n π.

Then, does the superconducting state break down? Actually, no. Cooper pairs prefer to preserve the superconducting state by producing a supercurrent. The supercurrent flows near the inner wall of the hollow torus so that the total magnetic flux, i.e.
Φ plus the magnetic field induced by the supercurrent, is equal to n (h/2e). Although there seem to be many possible choices for arriving at a quantized flux, only the quantized flux closest to Φ is automatically selected to determine n.

This flux quantization occurs only when a magnetic field is completely surrounded by a superconductor. Conversely, observation of the flux quantization in the present case is confirmation that the Nb layer becomes superconducting, and that a magnetic flux is completely surrounded by the superconductor.

Since a magnetic flux of h/2e produces an electron phase shift of π, the relative phase shift inside and outside a toroidal sample is 0 or π depending upon whether an even or odd number of flux quanta are trapped. This explains the observed phase shift quantization in Fig. 5.26.

5.5.9 Estimation of Overlap between Electron and Magnetic Field

It is natural to wonder whether the overlap between an electron beam and a magnetic field completely vanishes in this experiment. Unfortunately, no experiment can realize a completely zero overlap between the two. The tail of the electron wave function extends to the end of the world when infinitely small amounts are taken into account. A magnetic field cannot completely be shielded within a superconductor, but only decreased along the lines exp (-t/λ), where t and λ are respectively the thickness and penetration depth of the superconductor.

In the present experiment, the portion of the incident electron beam coherently reaching the magnetic field inside the toroidal sample was estimated to be of the order of 10^{-6} at the most. This is because a 150kV electron beam had to penetrate through the Cu (1000Å) and Nb (3000Å) layers to reach the magnetic field (see Osakabe et al. [205]). Sufficient prevention of electron penetration was also confirmed by experimental results which showed that a change in Cu-layer thickness from 500 to 2000Å had no effect on the interference fringes around the quantized magnetic flux.

Leakage flux from the magnet, which might influence the electron waves, could be estimated to be far less than $(1/200)(h/e)$. This was because any small amount of leakage flux $\sim(1/20)(h/e)$ from the toroidal magnet was further shielded by the Meissner effect of the superconducting niobium layer (thickness: 3000Å, penetration depth: 1000Å).

The degree of overlap between the electron beam and the magnetic field was thus estimated to be sufficiently small, even if not exactly zero. Since the existence or nonexistence of the AB effect should be verified experimentally, the only way to determine it is to believe in the continuity of physical phenomena between negligibly small and absolutely zero overlap.

5.5.10 Implications of Experiment

The existence of the AB effect was confirmed in this experiment where there was no overlap between an electron wave and magnetic fields. The electron wave could only have been affected by gauge fields (vector potentials) in field-free regions.

This experiment can also be considered to negate Liebowitz's classical interpretation [34] of the AB effect, which states that an electron passing by a solenoid receives a force resulting from the overlap of the electron's magnetic field with the solenoid's magnetic field (see Fig. 4.2). In the present experiment, a super-conducting barrier was built between the electron and magnetic field to prevent any overlap, just as in the experiment proposed by Erlichson [40]. Therefore, such a classical force cannot explain the AB phase shift.

Without a doubt, the classical interpretation is incompatible with the present experimental conditions. That is, there was no chance for two electrons to exist in the experimental apparatus at one time, and the accumulation of successive single electrons built up the electron interference pattern. This procedure can be seen in a time series of movie frames (see Fig. 5.29) demonstrating the single-electron buildup of the interference pattern [206].

In conclusion, the AB effect is purely quantum-mechanical.

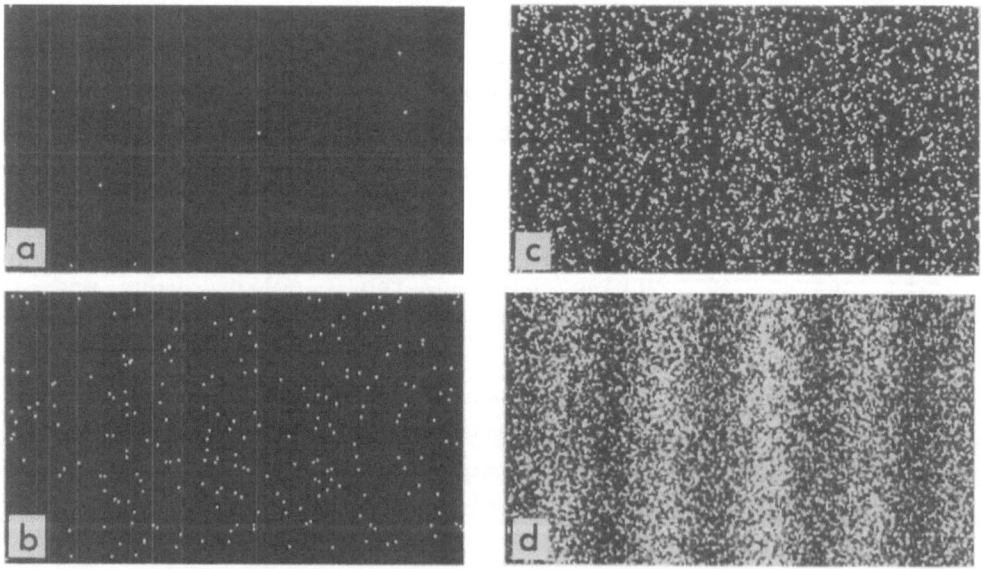

Fig. 5.29 Single-electron buildup of an electron interference pattern: (a) Number of electrons = 10, (b) number of electrons = 100, (c) number of electrons = 3,000, and (d) number of electrons = 70,000. The average number of electrons in the apparatus at one time is approximately 10^{-5}. Therefore, we must conclude that a single electron passes through both sides of the electron biprism and forms the probability amplitude of the biprism interference pattern.

6. CONCLUSIONS

As we have seen, the Aharonov-Bohm effect has been the subject of ongoing discussions - and controversy - for almost three decades. However, the fact that success has recently been achieved in confirming the AB effect experimentally now removes any doubt about the existence of the effect. As was discussed in Part One, the existence of the effect is further supported by a growing body of theoretical studies. One clear and lucid example is the paper by Yang [207] presented at the International Symposium on the Foundations of Quantum Mechanics (ISQM) in 1983.

It is perhaps natural to ponder the reasons why the controversy has been so longstanding. Determining these reasons may help to bring the features of the AB effect into relief from yet another angle. Before closing, I'd thus like to try to pull these reasons together. I see three reasons here.

To start with, one reason driving people against the AB effect is undoubtedly the fact that it appears so obviously inconceivable within the logic of classical physics. Secondly, the AB effect is not only unusual, but also physically significant. That is, it is directly related to such fundamentals of quantum mechanics as the single-valuedness of the wave function, the reality of vector potentials, and the locality of physical effects. What is more, the AB effect directly represents the gauge principle which is now playing an increasingly important role in unifying fundamental inter-actions in nature. The AB effect is regarded as experimental evidence for the physical reality of gauge fields. And thirdly, it has been extraordinarily difficult to experimentally confirm the theoretical predictions regarding the AB effect as experimental setups need to be made on an extremely small scale. In fact, it was only thanks to recent advanced techniques of electron holography and microlithography that a close-to-perfect experiment could be carried out.

Actually cases where recent developments in advanced technology have opened the way to the experimental verification of fundamental physical phenomena previously inaccessible are not limited to this example of the AB effect. Another example is the experimental verification of the AB effect for electrons in a normal metal, as carried out by Webb et al. [208]. In their experiment, an extremely thin gold ring only 300Å thick was fabricated using advanced lithographic techniques. AB resistance oscillation could not be detected until such a tiny sample was employed. Previously, magnetic flux enclosed by a loop in the ring could not be determined with precision much higher than h/2e due to the thick rings employed.

Yet another example is a neutron interferometer. Recently, several decisive experiments confirming quantum-mechanical effects have been performed by neutron interferometry (see Klein and Werner [209], Greenberger [210], Zeilinger [211], and Rauch [212]). One reason why this is such a recent development is that it was not possible to put neutron interferometry to practical use until a very large silicon crystal without imperfections could be utilized as the interferometer.

As Professor S. Nakajima of Tokai University stressed in his opening address at the Second ISQM (1986) [213], the future will most likely see advanced technology and fundamental physics coming closer and closer together. To cite a concrete example in my field, we can expect breakthroughs in the ease of viewing interference phenomena for electron wave functions due to further developments in both coherent electron beams and electron holography. In ways not yet envisioned, microlithography might be utilized to realize other experiments which have been relegated to a purely mental sphere (Gedankenexperimente). In this way we can probe previously inaccesible fundamental aspects of quantum mechanics.

REFERENCES

[1]   Y. Aharonov and D. Bohm: Phys. Rev. 115 (1959) 485.

[2]   T. T. Wu and C. N. Yang: Phys. Rev. D 12 (1975) 3845.

[3]   P. Bocchieri and A. Loinger: Nuovo Cimento 47A (1978) 475.

[4]   W. Ehrenberg and R. W. Siday: Proc. Phys. Soc. London B62(1949) 8.

[5]   F. Lenz: Phys. Bl. 18 (1962) 305; see also F. Lenz: Naturwiss. 48 (1961) 82.

[6]   D. H. Kobe: Ann. Phys. (N.Y.) 123 (1979) 381.

[7]   L. Marton: Science 118 (1953) 470.

[8]   F. G. Werner and D. R. Brill: Phys. Rev. Lett. 4 (1960) 344.

[9]   R. G. Chambers: Phys. Rev. Lett. 5 (1960) 3.

[10]   H. A. Fowler, L. Marton, J. A. Simpson and J. A. Suddeth: J. Appl. Phys. 32 (1961) 1153.

[11]   G. Möllenstedt and W. Bayh: Phys. Bl. 18 (1962) 299.

[12]   H. Boersch, H. Hamisch and K. Grohmann: Z. Phys. 169 (1962) 263.

[13]   R. C. Jaklevic, J. Lambe, J. E. Mercereau and A. H. Silver: Phys. Rev. A 140 (1965) 1628.

[14]   G. Matteucci and G. Pozzi: Am. J. Phys. 46 (1978) 619.

[15]   S. Olariu and I. I. Popescu: Rev. Mod. Phys. 57 (1985) 339.

[16]   G. Möllenstedt and H. Dücker: Z. Phys. 145 (1956) 377.

[17]   W. Bayh: Z. Phys. 169 (1962) 492.

[18]   G. Möllenstedt and W. Bayh: Naturwiss. 4 (1962) 81.

[19]   Y. Aharonov and D. Bohm: Phys. Rev. 123 (1961) 1511.

[20]   G. Möllenstedt and E. Krimmel: Z. Angew. Phys. 16 (1963) 121.

[21]   W. H. Furray and N. F. Ramsey: Phys. Rev. 118 (1960) 623.

[22]   H. Wegener: Z. Phys. 159 (1960) 243.

[23]   H. E. Mitler: Phys. Rev. 124 (1961) 940.

[24] R. P. Feynman, R. B. Leighton and M. Sands: in The Feynman Lectures in Physics (Addison-Wesley, Reading MA, 1964) Vol. 2, p.15.

[25] M. Peshkin, I. Talmi and L. J. Tassie: Ann. Phys. 16 (1961) 426.

[26] L. J. Tassie and M. Peshkin: Ann. Phys. (N.Y.) 16 (1961) 177.

[27] P. D. Noerdlinger: Nuovo Cimento 23 (1962) 158.

[28] B. S. DeWitt: Phys. Rev. 125 (1962) 2189.

[29] F. J. Belinfante: Phys. Rev. 128 (1962) 2832.

[30] E. L. Feinberg: Sov. Phys. Usp. 5 (1963) 753.

[31] G. T. Trammel: Phys. Rev. B 134 (1964) 1183.

[32] Y. Aharonov and D. Bohm: Phys. Rev. 125 (1962) 2192.

[33] A. Aharonov and D. Bohm: Phys. Rev. 130 (1963) 1625.

[34] B. Liebowitz: Nuovo Cimento 38 (1965) 932.

[35] P. Hraskó: Nuovo Cimento B44 (1966) 452.

[36] B. Liebowitz: Nuovo Cimento B46 (1966) 124.

[37] E. Kasper: Z. Phys. 196 (1966) 415.

[38] E. Kasper: Z. Phys. 207 (1967) 24.

[39] B. Liebowitz: Z. Phys. 207 (1967) 20.

[40] H. Erlichson: Am. J. Phys. 38 (1970) 162.

[41] B. S. Deaver, Jr., and W. M. Fairbank: Phys. Rev. Lett. 7 (1961) 43.

[42] B. Lischke: Z. Phys. 239 (1970) 360.

[43] H. Wahl: Optik 28 (1968/1969) 417.

[44] T. H. Boyer: Phys. Rev. D 8 (1973) 1667.

[45] T. H. Boyer: Phys. Rev. D 8 (1973) 1679.

[46] C. N. Yang and R. L. Mills: Phys. Rev. 96 (1954) 191.

[47] R. Utiyama: Phys. Rev. 101 (1956) 1957.

[48] S. Weinberg: Phys. Rev. Lett. 19 (1967) 1264.

[49] A. Salam: Proc. of the 8th Int. Symp. on Elementary Particle Therory, edited by N. Svartholm (Almqvist & Wilksells, Stockholm, 1968) p.367.

[50] H. J. Bernstein and S. V. Philips: Scientific American 245 (July, 1981) 95.

[51] P. Tourrenc: Phys. Rev. D 16 (1977) 3421.

[52] A. Zeilinger, M. Horn and C. G. Shull: Proc. Int. Symp. on Foundations of Quantum Mechanics, Tokyo, 1983, edited by S. Kamefuchi et al. (Physical Society of Japan, Tokyo, 1984) p.289.

[53] A. Zeilinger: J. de Phys. 45 (1984) C3, 213.

[54] L. C. L. Botelho and J. C. de Mello: J. Phys. A18 (1985) 2633.

[55] P. A. Horváthy: Phys. Rev. D 33 (1986) 407.

[56] R. Sundrum and L. J. Tassie: J. Math. Phys. 27 (1986) 1566.

[57] R. P. Feynman: Rev. Mod. Phys. 20 (1948) 367.

[58] D. Wisnivesky and Y. Aharonov: Ann. Phys. 45 (1967) 479.

[59] J. S. Dowker: Nuovo Cimento B52 (1967) 129.

[60] G. Papini: Nuovo Cimento B52 (1967) 136.

[61] D. Greenberger: Ann. Phys. (N.Y.) 47 (1968) 116.

[62] K. Krauss: Ann. Phys. (N.Y.) 50 (1968) 102.

[63] A. W. Overhauser and R. Collela: Phys. Rev. Lett. 33 (1974) 1237.

[64] J. Anandan: Phys. Rev. D 15 (1977) 1448.

[65] J. Anandan: Nuovo Cimento A 53 (1979) 221.

[66] L. H. Ford and A. Vilenkin: J. Phys. A 14 (1981) 2353.

[67] J. Stachel: Phys. Rev. D 26 (1982) 1281.

[68] C. J. C. Burges: Phys. Rev. D 32 (1985) 504.

[69] J. A. Ferrari and J. Griego: Nuovo Cimento 96B (1986) 41.

[70] R. Parthasarathy, G. Rajasekaran and R. Vasudevan: Class. Quantum Grav. 3 (1986) 425.

[71] V. B. Bezerra: Phys. Rev. D 35 (1987) 2031.

[72] Y. Aharonov and M. Vardi: Phys. Rev. D 20 (1979) 3213.

[73] P. C. Naik: Pramana-J. Phys. 27 (1986) 629.

[74] D. M. Greenberger, D. K. Atwood, J. Arthur, C. G. Shull and M. Schlenker: Phys. Rev. Lett. 47 (1981) 751.

[75] Y. Aharonov and A. Casher: Phys. Rev. Lett. 53 (1984) 319.

[76]   A. G. Klein: Physica 137B (1986) 230.

[77]   P. Bocchieri, A. Loinger and G. Siragusa: Nuovo Cimento 51A
        (1979) 1.

[78]   D. Home and S. Sengupta: Am. J. Phys. 51 (1983) 942.

[79]   P. Bocchieri, A. Loinger and G. Siragusa: Nuovo Cimento 56A
        (1980) 55.

[80]   W. C. Henneberger: J. Math. Phys. 22 (1981) 116; see also
        W. C. Henneberger: Phys. Rev. A 22 (1980) 1383;
        W. C. Henneberger: Phys. Rev. Lett. 52 (1984) 573.

[81]   J. Q. Liang: Nuovo Cimento 92B (1986) 167.

[82]   D. H. Kobe and J. Q. Liang: Phys. Lett. A118 (1986) 475.

[83]   U. Klein: Lett. Nuovo Cimento 25 (1979) 33; see also
        U. Klein: Acta Phys. Austriaca 52 (1980) 269.

[84]   A. Zeilinger: Lett. Nuovo Cimento 25 (1979) 333.

[85]   D. Bohm and B. J. Hiley: Nuovo Cimento 52A (1979) 295.

[86]   J. A. Mignaco and C. A. Novaes: Lett. Nuovo Cimento 26
        (1979) 453.

[87]   M. Bawin and A. Burnel: Lett. Nuovo Cimento 27 (1980) 4.

[88]   P. Bocchieri and A. Loinger: Lett. Nuovo Cimento 25 (1979)
        476.

[89]   W. C. Henneberger: Am. J. Phys. 52 (1984) 375.

[90]   A. Burnel and V. Reekmans: Am. J. Phys. 53 (1985) 777.

[91]   D. Home: Am. J. Phys. 53 (1985) 778.

[92]   E. Madelung: Z. Phys. 40 (1926) 322.

[93]   D. Bohm: Phys. Rev. 85 (1952) 166; ibid., 85 (1952) 180.

[94]   T. Takabayasi: Prog. Theor. Phys. 8 (1952) 143; ibid., 9
        (1953) 187.

[95]   F. Strocchi and A. S. Wightman: J. Math. Phys. 15 (1974)
        2198.

[96]   T. Takabayasi: Prog. Theor. Phys. 69 (1983) 1323.

[97]   T. Takabayasi: Proc. Int. Symp. on Foundations of Quantum
        Mechanics, Tokyo, 1983, edited by S. Kamefuchi et al.
        (Physical Society of Japan, Tokyo, 1984) p.44.

[98]   L. Jánossy: Acta Physica, Acad. Scien. Hung. 29 (1970) 419.

[99]   C. Casati and I. Guarneri: Phys. Rev. Lett. 42 (1979) 1579.

[100] K. Wódkiewicz: Phys. Rev. A 29 (1984) 1527.

[101] A. Y. Shiekh: Ann. Phys. (NY) 166 (1986) 299.

[102] J. Q. Liang and Xiu-Xiang Ding: Phys. Lett. A119 (1987) 325.

[103] V. F. Weisskopf: in Lectures in Theoretical Physics, edited by W. E. Brittin, B. W. Downs and J. Downs (Interscience, New York, 1961) p.63.

[104] M. Peshkin: Phys. Rep. 80 (1981) 375.

[105] F. Wilczek: Phys. Rev. Lett. 48 (1982) 1144.

[106] D. H. Kobe: J. Phys. A 15 (1982) L543.

[107] G. A. Goldin, R. Menikoff and D. H. Sharp: J. Math. Phys. 22 (1981) 1664.

[108] P. Bocchieri and A. Loinger: Nuovo Cimento 66A(1981) 164.

[109] P. Frolov and V. D. Skarzhinsky: Nuovo Cimento 76 (1983) 35.

[110] P. Bocchieri and A. Loinger: Lett. Nuovo Cimento 39 (1984) 148.

[111] S. M. Roy and V. Singh: Nuovo Cimento 79A (1984) 391.

[112] V. L. Lyuboshits and Ya. A. Smorodinskii: Sov. Phys. JETP 48 (1978) 19.

[113] M. Kretzschmar: Z. Phys. 185 (1965) 84.

[114] E. Corinaldesi and F. Rafeli: Am. J. Phys. 46 (1978) 1185.

[115] M. V. Berry: Eur. J. Phys. 1 (1980) 240; see also M. V. Berry, R. G. Chambers, M. D. Large, C. Upstill and J. C. Walmsley: Eur. J. Phys. 1 (1980) 154.

[116] M. Bawin and A. Burnell: J. Phys. A 16 (1983) 2173.

[117] S. N. M. Ruijsenaars: Ann. Phys. (NY) 146 (1983) 1.

[118] Y. Aharonov, C. K. Au, E. C. Lerner and J. Q. Liang: Phys. Rev. D 29 (1984) 2396.

[119] B. Nagel: Phys. Rev. D 32 (1985) 3328; see also B. Nagel: Int. J. Quantum Chemistry 19 (1986) 741.

[120] R. A. Brown: J. Phy. A 18 (1985) 2497.

[121] K. Kawamura: Z. Phys. 29 (1978) 101; ibid., B 30 (1978)1.

[122] H. Boersch, K. Grohmann, H. Hamisch, B. Lischke and D. Wohlleben: Lett. Nuovo Cimento 30 (1981) 257.

[123] P. Bocchieri and A. Loinger: Lett. Nuovo Cimento 30 (1981) 449.

[124]   S. M. Roy: Phys. Rev. Lett. 44 (1980) 111.

[125]   U. Klein: Phys. Rev. D 23 (1981) 1463.

[126]   H. J. Lipkin: Phys. Rev. D 23 (1981) 1466.

[127]   D. M. Greenberger: Phys. Rev. D 23 (1981) 1460.

[128]   M. Peshkin: Phys. Rev. A 23 (1981) 360.

[129]   C. G. Kuper: Phys. Lett. 794 (1980) 413.

[130]   M. Babiker and R. Loudon: J. Phys. A 17 (1984) 2973.

[131]   A. S. Goldhaber: Phys. Rev. Lett. 49 (1982) 905.

[132]   H. J. Lipkin and M. Peshkin: Phys. Lett. 118 B (1982) 385.

[133]   R. Jackiw and S. N. Redlich: Phys. Rev. Lett. 50 (1983) 555.

[134]   M. P. Silverman: Lett. Nuovo Cimento 41 (1984) 509.

[135]   M. P. Silverman: Phys. Rev. Lett. 51 (1983) 1927; see also
        M. P. Silverman: Phys. Rev. D 29 (1984) 2404.

[136]   P. A. Horváthy: Phys. Rev. A 31 (1985) 1151.

[137]   J. Q. Liang: Phys. Rev. Lett. 53 (1984) 859.

[138]   Morandi: Lett. Nuovo Cimento 39 (1984) 312.

[139]   A. Tonomura, T. Matsuda, R. Suzuki, A. Fukuhara,
        N. Osakabe, H. Umezaki, J. Endo, K. Shinagawa,
        Y. Sugita and H. Fujiwara: Phys. Rev. Lett. 48 (1982) 1443.

[140]   L. S. Schulman: J. Math. Phys. 12 (1971) 304.

[141]   A. Inomata and V. A. Singh: J. Math. Phys. 19 (1978) 2318.

[142]   C. C. Bernido and A. Inomata: Phys. Lett. 77A (1980) 394.

[143]   C. C. Bernido and A. Inomata: J. Math. Phys. 22 (1981) 715.

[144]   B. S. Deaver Jr. and G. B. Donaldson: Phys. Lett. 89A (1982)
        178.

[145]   C. Gerry and V. A. Singh: Phys. Lett. 92A (1982) 11.

[146]   A. Inomata: Phys. Lett. 95A (1983) 176.

[147]   C. Morandi and E. Menossi: Eur. J. Phys. 5 (1984) 49.

[148]   Y. Ohnuki: Proc. Int. Symp. on Foundations of Quantum
        Mechanics, Tokyo, 1986, edited by M. Namiki et al. (Physical
        Society of Japan, Tokyo, 1987) p.117.

[149]   A. Inomata: Proc. Int. Symp. on Foundations of Quantum
        Mechanics, Tokyo, 1986, edited by M. Namiki et al. (Physical
        Society of Japan, Tokyo, 1987) p.132.

[150]   M. V. Berry: Proc. R. Soc. Lond. A392 (1984) 45.

[151]   B. Simon: Phys. Rev. Lett. 51 (1983) 2167.

[152]   A. Tomita and R. Y. Chiao: Phys. Rev. Lett. 57 (1986) 937.

[153]   G. Delacretaz, E. R. Grant, R. L. Whetten, L. Wöste and
        J. W. Zwanziger: Phys. Rev. Lett. 56 (1986) 2598.

[154]   J. Anandan and L. Stodolsky: Phys. Rev. D. 35 (1987) 2597.

[155]   Y. Aharonov and J. Anandan: Phys. Rev. Lett. 58 (1987) 1593.

[156]   R. Doll and M. Näbauer: Phys. Rev. Lett. 7 (1961) 51.

[157]   F. London, in Superfluids (John Wiley & Sons, Inc., New York,
        1950) p.152.

[158]   L. Onsager: Proc. Int. Conf. on Theoretical Physics, Kyoto
        and Tokyo, 1953 (Science Council of Japan, Tokyo, 1954)
        p.935.

[159]   N. Byers and C. N. Yang: Phys. Rev. Lett. 7 (1961) 46.

[160]   M. Peshkin: Phys. Rev. 132 (1963) 14.

[161]   J. Bardeen: Phys. Rev. Lett. 7 (1961) 162.

[162]   B. Lischke: Phys. Rev. Lett. 22 (1969) 1366.

[163]   O. Costa de Beauregard and J. M. Vigoureux: Lett. Nuovo
        Cimento 33 (1982) 79.

[164]   H. Jehle: Proc. Monopole Meeting, Trieste, 1981, eds.
        N. S. Craigie, P. Goddard and W. Nahm (World Scientific,
        Singapore, 1982) p.411.

[165]   E. J. Post: Phys. Lett. 92 (1982) 224.

[166]   G. Kunstatter: Can. J. Phys. 62 (1984) 737.

[167]   Y. Ne'eman: Foundations of Physics 16 (1986) 361.

[168]   G. 't Hooft: Physica Scripta 25 (1982) 133.

[169]   R. A. Ferrell and J. J. Hopfield: Physics 1 (1964) 1.

[170]   E. Lubkin: Am. J. Phys. 39 (1971) 94.

[171]   P. A. M. Dirac: Proc. R. Soc. London A133 (1931) 60.

[172]   P. Goddard and P. Olive: Rep. Prog. Phys. 41 (1978) 1357.

[173]   S. M. Roy and V. Singh: Phys. Rev. Lett. 51 (1983) 2069.

[174]   A. O. Barut and R. Wilson: Ann. Phys. 164 (1985) 223.

[175]   M. H. Saffouri: Nuovo Cimento 96A (1986) 1.

[176]   A. O. Barut: J. Phys. A 11 (1978) 2073.

[177]   Y. Aharonov: Proc. Int. Symp. on Foundations of Quantum
        Mechanics, Tokyo, 1983, edited by S. Kamefuchi et al.
        (Physical Society of Japan, Tokyo, 1984) p.10.

[178]   B. I. Spasskii and A. V. Moskovskii: Sov. Phys. Usp. 27
        (1984) 273.

[179]   N. G. Van Kampen: Phys. Lett. 106A (1984) 5.

[180]   T. Troudet: Phys. Lett. 111 (1985) 274.

[181]   J. J. Tassie: Phys. Lett. 5 (1963) 43.

[182]   H. J. Rothe: Nuovo Cimento 62A (1981) 54.

[183]   A. Tonomura: in Progress in Optics, Vol. 23, edited by
        E. Wolf (North-Holland, Amsterdam, 1986) p.183.

[184]   A. Tonomura, H. Umezaki, T. Matsuda, N. Osakabe, J. Endo and
        Y. Sugita: Proc. Int. Symp. on Foundations of Quantum
        Mechanics, Tokyo, 1983, edited by S. Kamefuchi et al.
        (Physical Society of Japan, Tokyo, 1984) p.20.

[185]   A. Tonomura: Rev. Mod. Phys. 59 (1987) 639.

[186]   D. Gabor: Proc. Roy. Soc. London A197 (1949) 454.

[187]   A. Tonomura, T. Matsuda, T. Kawasaki, J. Endo and
        N. Osakabe: Phys. Rev. Lett. 56 (1985) 792.

[188]   P. Bocchieri, A. Loinger and G. Siragusa: Lett. Nuovo
        Cimento 35 (1982) 370.

[189]   H. Miyazawa: Proc. 10th Hawaii Conf. on High Energy Physics,
        Hawaii, 1985, p.441.

[190]   E. Shrödinger: Ann. Physik 32 (1938) 49.

[191]   W. Pauli: Helv. Phys. Acta 12 (1939) 147.

[192]   J. M. Blatt and V. F. Weisskopf: in Theoretical Nuclear
        Physics (John Wiley & Sons. Inc., New York, 1952) p.783, 787.

[193]   E. Merzbacher: Am. J. Phys. 30 (1962) 237.

[194]   M. Kretzchmar: Z. Phys. 185 (1965) 73.

[195]   M. P. Silverman: Lett. Nuovo Cimento 42 (1985) 37.

[196]   E. Comay: Physica 135 A (1986) 281.

[197]   A. Tonomura, T. Matsuda, J. Endo, T. Arii and K. Mihama:
        Phys. Rev. B. 34 (1986) 3397.

[198]   I. R. Walker: Phys. Rev. B 33 (1986) 5028.

[199]   G. Möllenstedt, H. Schmid and H. Lichte: Proc. 10th Int.
        Conf. on Electron Microscopy (Hamburg, 1982) Vol.1, p.433.

[200]   A. Tonomura, H. Umezaki, T. Matsuda, N. Osakabe, J. Endo and
        Y. Sugita: Phys. Rev. Lett. 51 (1983) 331.

[201]   G. Matteucci and G. Pozzi: Phys. Rev. Lett. 54 (1985) 2469.

[202]   H. Schmid: Proc. the 8th European Cong. on Electron
        Microscopy, Budapest, 1984, edited by Á. Csanády et al.
        (Program Committee, Budapest, 1984) p.285.

[203]   A. Tonomura, N. Osakabe, T. Matsuda, T. Kawasaki, J. Endo, S.
        Yano and H. Yamada: Phys. Rev. Lett 56 (1986) 792.

[204]   A. Tonomura, S. Yano, N. Osakabe, T. Matsuda, H. Yamada, T.
        Kawasaki and J. Endo: Proc. Int. Symp. on Foundations of
        Quantum Mechanics, Tokyo, 1986, edited by M. Namiki et al.
        (Physical Society of Japan, Tokyo, 1987) p.97.

[205]   N. Osakabe, T. Matsuda, T. Kawasaki, J. Endo, A. Tonomura, S.
        Yano and H. Yamada: Phys. Rev. A 34 (1986) 815.

[206]   A. Tonomura, J. Endo, T. Matsuda, T. Kawasaki and H. Ezawa:
        Am. J. Phys. 57 (1989) 117.

[207]   C. N. Yang: Proc. Int. Symp. on Foundations of Quantum
        Mechanics, Tokyo, 1983, edited by S. Kamefuchi et al.
        (Physical Society of Japan, Tokyo, 1984) p.5.
        see also two papers: M. M. Nieto: Phys. Rev. A 29 (1984)
        3413; M. Berry: Proc. Workshop on Fundamental Aspects of
        Quantum Theory, Como, 1985, edited by V. Gorini and A.
        Frigerio (Plenum, New York, 1986) p.319.

[208]   R. A. Webb, S. Washburn, C. P. Umbach and R. B. Laibowitz:
        Phys. Rev. Lett. 54 (1985) 2696.

[209]   A. G. Klein and A. Werner: Rep. Prog. in Phys. 46 (1983)
        259.

[210]   D. M. Greenberger: Rev. Mod. Phys. 55 (1983) 875.

[211]   A. Zeilinger: Proc. Conf. on New Techniques and Ideas in
        Quantum Measurement Theory, New York City, 1986, edited by D.
        M. Greenberger (New York Academy of Science, New York, 1986)
        p.469.

[212]   H. Rauch: Proc. Int. Symp. on Foundations of Quantum
        Mechanics, Tokyo, 1986, edited by M. Namiki et al. (Physical
        Society of Japan, Tokyo, 1987) p.3.

[213]   S. Nakajima: Proc. Int. Symp. on Foundations of Quantum
        Mechanics, Tokyo, 1986, edited by M. Namiki et al. (Physical
        Society of Japan, Tokyo, 1987) p.1.

The explanation of the AB effect can be found in the following textbooks.

R. P. Feynman, R. B. Leighton and M. Sands: "The Feynman Lectures on Physics" (Addison-Wesley, Reading, Mass., 1964) pp. II.15.8-14.

G. Baym: "Lectures on Quantum Mechanics" (Benjamin, Reading, Mass. 1973) pp. 77-82.

A. Shadowitz: "The Electromagnetic Field" (McGraw-Hill Kogakusha, Tokyo, 1975) pp. 517-522.

J. L. Lopes: "Gauge Field Theories" (Pergamon, Oxford, 1981) pp. 98-101.

C. Quigg: "Gauge theories of the strong, weak, and electromagnetic interactions" (Benjamin/Cummings, Reading, Mass., 1983) pp. 43-45, p. 52.

J. J. Sakurai: "Modern Quantum Mechanics" (Benjamin/Cummings, Menlo Park, 1985) pp. 136-139.

APPENDIX

Many papers concerning the AB effect can be found in Conference Proceedings. Several are listed below for readers' convenience.

(1) Proceeding of the International Symposium on Foundations of Quantum Mechanics in the light of New Technology, Tokyo, 1983, edited by S. Kamefuchi, H. Ezawa, Y. Murayama, M. Namiki, S. Nomura, Y. Ohnuki and T. Yajima (Physical Society of Japan, Tokyo, 1984)

C. N. Yang   :"Gauge Fields, Electromagnetism and the Bohm-Aharonov Effect" .... p. 5

Y. Aharonov   :"Non-Local Phenomena and the Aharonov-Bohm Effect" p.10

A. Tonomura, H. Umezaki, T. Matsuda, N. Osakabe, J. Endo and Y. Sugita: "Electron Holography, Aharonov-Bohm Effect and Flux Quantization"...................... p.20

H. Lichte   :"Coherent Electron Optical Experiments Using an Electron Mirror" p.29

G. Matteucci, G. Missiroli, M. Porrini and G. Pozzi: "Recent Observations on a New Electrostatic Phase-Shifting Effect" p.39

T. Takabayasi: "Hydrodynamical Quantization and Spin" p.44

M. Peshkin   :"Aharonov-Bohm Effect: Influence of the Electron's Field, and Speculations on the Possibility of Shielding"................. p.56

<u>Post-Symposium Discussion-Meeting on the AB Effect</u> (p.367 - 368)

H. Lichte, G. Möllenstedt and H. Schmidt: "Application of an Electron Interferometer with Widely Separated Partial Waves"

G. Pozzi   : "Recent Results on a New Electrostatic Phase Shifting Effect"

Y. Aharonov   : "Dynamical Approach to AB Effect and New Effects Related to AB Effect"

O. Costa de Beauregard: "Flux Quantization in Autistic Magnet?"

M. Peshkin   : "Locality/Causality in the AB Effect"

J. S. Anandan: "Electron Interference in the Presence of the Shiff-Barnhill Field"

J. A. Wheeler: "Berry's Angle and AB Effect"

H. Miyazawa : "Multivalued Wave Functions"

S. Iida   : "The New Frame in Physics and the AB Effect"

M. A. Horne and A. Zeilinger:
          "AB Effects and Electric Charge-Magnetic Monopole
          Symmetry"

K. Kawamura : "Scattering Theoretical Solution of AB Equation"

K. Yasue   : "AB Effect and Inertial Force Effects in Stochastic
          Quantization -A Stochastic Variational Approach"

(2) Proceeding of the Conference on Fundamental Questions in
     Quantum Mechanics, Albany, 1984, edited by L. M. Roth and A.
     Inomata (Gordon and Breach, New York, 1986)

A. Tonomura : "Experimental Confirmation of the Aharonov-Bohm
          Effect by Electron Holography" p.169

M. P. Silverman:
          "Angular Momentum and Rotational Properties
          of a Charged Particle Orbiting a Magnetic
          Flux Tube"................................... p.177

M. D. Semon and J. R. Taylor:
          "A Dynamical Formulation of the Aharonov-Bohm
          Effect"...................................... p.19

C. C. Gerry and A. Inomata:
          "Topological Effects in Quantum Mechanics"... p.199

(3) Proceeding of the Workshop on Fundamental Aspects of Quantum
     Theory, Como, 1985, edited by V. Gorini and A. Frigerio
     (Plenum, New York, 1986)

A. Zeilinger: "Generalized Aharonov-Bohm Experiments with
          Neutrons" p.311

M. Berry   : "The Aharonov-Bohm Effect is Real Physics
          not Ideal Physics"........................... p.319

A. Loinger   : "Again About an Old Stuff:
          The Aharonov-Bohm Effect" p.321

P. Bocchieri: "Against the Existence of the
          Aharonov-Bohm Effect" p.325

M. Peshkin   : "Theories Without AB Effect Misrepresent the
          Dynamics of the Electromagentic Field"....... p.329

B. Nagel   : "Some Cases of the Aharonov-Bohm Effect" p.335

Lecture Notes in Mathematics

Lecture Notes in Physics